*The clapper bridge at Postbridge on Dartmoor*

# MEDIEVAL BRIDGES

# Mike Salter

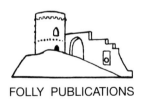

FOLLY PUBLICATIONS

## ACKNOWLEDGEMENTS

Firstly, a big thanks to my partner Val Webb, who helped with research, driving hire-cars and proof reading, and who took the photographs of the bridges at Aysgarth, Bowbridge, Kildwick, Lincoln, Otley, Rotherham, Wensley and the Uppermill stepping-stones. The author took all the other modern pictures during 2013-15, and also drew the plans and the maps which appear at the end of each gazetteer. Many thanks are also due to Bernard Mutton, who spent two days driving the author around many of Scotland's bridges, and to Hilary Rodwell and Rosemary Hollins, each of whom drove us to several bridges and provided accommodation. Others who provided accommo-dation for a night during our travels to see bridges were Jenny Harper, Ian Morris, Milly Morris and Dave Isaacson and Peter and Elaine Ryder.

## CRITERIA FOR INCLUSION OF BRIDGES

The aim has been to include all surviving bridges in Britain where at least some part of an existing structure is thought to probably predate the Reformation of the mid 16th century. In Britain the medieval period is usually taken as going up to the end of the 15th century. However the connection between the church and the building and main-tenance of bridges and the provision of chapels on or beside them was so strong dur-ing the whole of the period up to the time of the Reformation that going up to the mid 16th century makes sense. For England and Wales, John Leland's Itinerary of c1535-42 happens to coincide with the Reformation period and provides us with much useful information about which bridges were then at least partly of either stone or brick.

Bridges or constituent parts of them can be difficult to date even to a particular century, let alone to a specific point within a century. Most late 16th and 17th century bridges tended to continue with the general style, dimensions and building techniques of medieval bridges, so it is impossible to be precise about whether certain bridges are of before or after the stated cut-off point. The fact that the Reformation took effect a few years later in Scotland has also been taken into account although there are no bridges there assumed to date precisely from the critical period c1540-60. Many bridges which are generally regarded as 16th century have been included, unless there is a specific reason for assuming that nothing of the present fabric is likely to predate the 1560s. The oldest bridge excluded from the main gazetteers is that of 1563 at Brecon.

Bridges in this book mostly lie on public roads or rights-of-way, although in some cases little or nothing of the oldest parts can be seen from road level. In several instanc-es old features are only visible from the water or private land not crossed by paths.

Many bridges have multiple names. Place names have been generally been prefered for the gazetteer entry headers, with other names usually given in the entry.

## ABOUT THE AUTHOR

Mike Salter is 61 and has been a professional writer and publisher since he went on the Government Enterprise Allowance Scheme for unemployed people in 1988. He is particularly interested in the planning and layout of medieval buildings and has a huge collection of plans of churches and castles he has measured during tours (mostly by bi-cycle, motorcycle or on foot) throughout all parts of the British Isles since 1968. Wolver-hampton born and bred, Mike now lives in an old cottage beside the Malvern Hills. His other interests include walking, maps, railways, board games, various types of morris and folk dancing, mumming, playing percussion instruments and calling folk dances.

*The Welsh Bridge at Shrewsbury as it was pictured in the 1790s*

## CONTENTS

A map is located at the end of each gazetteer section

# INTRODUCTION

There is little evidence to suggest bridges were common in Britain before the Roman period. Archaeological evidence suggests that waterborne traffic was more important than overland routes in pre-Roman times, especially for heavy or bulky goods. Most rivers in their natural state before canalisation and drainage works carried out since Tudor times could be forded at various points except in spate conditions. Depths of water up to three feet for foot traffic and up to four feet for those on horseback seem to have been regarded as acceptable by those not laden with trade goods. Even the Medway and Thames were forded easily enough by Roman troops during the invasion of AD43. The Thames at London is known to have been wider and shallower in Roman times than it is today. Place names recording fords are common enough, as in the towns of Stafford, Stamford and Wallingford. Names such as Langford or Longford indicate a long ford, whilst Bradford suggests a broad width of safe crossing. Shalford is probably a corruption of shallow ford whilst the very common place-name of Stratford implies a ford on the line of a street, the name then usually only given to a Roman road. Fords commonly remained in use beside bridges, being used for stock movements and loads too wide or heavy for small bridges to bear, and there are still many fords on country lanes in England where the bridge alongside (whatever its date) is only suitable for pedestrian traffic. Fords remained extremely common until the mass production of concrete pipes made culverts more easy to construct in the early 20th century.

Small ferries also used to be very common and there are many instances of where dead-end lanes lead to opposite banks of a river, across which a ferry remained in use until motorised vehicles became the norm in the mid 20th century. Ferries often supplemented bridges, some of which, particularly in towns, could be very congested, and in any case medieval bridges were sometimes too dilapidated to be safely crossed. Ferries could be profitable and in some cases there was a consequent reluctance to repair a broken bridge, as discovered by a commission set up in 1365 to enquire into the delay in repairing a bridge at Newenden on the Kent-Sussex border. At Hampton near Evesham in Worcestershire a cable ferry survives in use.

Another alternative to a bridge was the provision of a row of stepping stones across a river, which would simply flow over the top of them in times of spate. Many examples still exist serving footpaths. One or two are likely to be of medieval origin because of their topographic context, such as the stones across the River Ogmore below Ogmore Castle near Bridgend. Stepping stones associated with weirs or industrial workings or which are fairly regular in size and shape, such as those across the River Wharfe near Bolton Priory, are likely to be more recent.

*The small packhorse bridge beside a ford in the village of Rearsby in Leicestershire*

*Uppermill: stepping stones
over the River Tame*

*Southern abutment of Roman bridge over River Tees at Piercebridge*

Long distance paths such as the Ridgeway over the Berkshire Downs and the Icknield Way extending out to East Anglia are generally thought of as perpetuating ancient trackways, gradually created by constant treading over the passage of time. Roads in the sense of engineered routeways with properly drained stone surfaces, more suitable for wheeled traffic, were long thought of as having been introduced by the Romans, who are also credited as builders of the first proper bridges. However evidence of older roadways has begun to turn up in excavations. Wooden walkways across the marshes of the Somerset Levels existed in late Mesolithic times, since sections of the Sweet Track between Westhay and Shapwick were made with timber cut down c3800BC. Excavations in advance of quarrying at Sharpstones in Shropshire discovered a 1st century BC road with a stone surface that was cambered ie sloped down from the centre. Grid patterns of metalled (stone surfaced) streets also existed in pre-Roman times at tribal centres at Danbury and Silchester in Hampshire.

The Romans are thought to have created 9,500 miles worth of new roads within the first hundred years of their occupation of Britain and the lines taken by over three quarters of these roads are now known, together with the locations of a hundred bridges.

Cornwall is the only English county lacking evidence of Roman roads, which were noted for having long straight sections. Their original purpose was military, allowing the rapid transit of troops, supplies and messages. Remnants exist of where Hadrian's Wall and its associated military road crossed rivers at Chesters and Corbridge. The other chief visitable remnant of a Roman bridge is at Piercebridge on the River Tees. Footings remain of the southern abutment and several piers with triangular cutwaters of what is generally thought to have been a timber-decked bridge. Parts of other Roman bridges have been excavated in London and at Aldwinkle in Northamptonshire.

*Clapper bridge at Austwick in North Yorkshire*

Some of the lesser Roman roads, especially those in upland areas, fell into disuse during the Anglo-Saxon period but other roads and their associated bridges and fords remained in use insofar as a lack of maintenance allowed. Bede mentions paved roads in the 8th century, but it does not appear that there was a bridge in use across the Thames at London before Southwark came into being in the late 9th century. Historical records of bridges during this period are few and the earliest new large scale water crossing after the Roman period known from excavations is a piled causeway of c700 linking Mersea Island with the Essex mainland.

Roman canals such as the Foss Dyke in Lincolnshire seem to have remained open since water travel was important during this period. One Late Saxon canal known to us is the cutting off of a loop of the River Thames by the monks of Abingdon Abbey in the 1050s. The most notable of later medieval canal schemes in Britain was the canalisation of the river Don in south Yorkshire. Several abbeys (eg Rievaulx) and castles are known to have each had a short length of canal connecting them with a river or lake to ease the transhipment of building materials.

New roads were created to link the burhs or fortified towns created by Saxon kings in the late 9th and early 10th century on the model of those of Charles the Bald in France. Little is known of these roads and still less of what arrangements were made for river crossings along these new routes. King Alfred is known to have built a bridge at Lyng in Somerset but more frequent references to bridges in charters only occur in the 10th century. The Anglo-Saxon Chronicle mentions the building of a bridge at Nottingham, and there were others at Chester, Stamford, Worcester and York. The few references suggesting possible stone structures during this period probably only refer to piers and abutments of masonry intended to support timber decking. All the burhs lay beside rivers, as at Wallingford and Wareham, and it appears that timber bridges were not only created as crossings controlled by these fortified places, but also as a means of controlling water-borne Viking raids. Excavations in Tamworth have revealed an abutment of a 10th century timber bridge and what is thought to have been another was found next to Roman bridge foundations at Staines in Surrey (formerly Middlesex). Parts of an 8th or 9th century trestle bridge have also been found at Oxford.

Cambridge takes its name from a bridge which existed by the 9th century. In c1000 the Roman bridge in London was replaced, timbers from the new bridge found reused in the foundations of the later bridge having been felled c987-1032. The last major Viking invasion of England was crushed by King Harold in a battle fought beside the bridge at Stamford in East Yorkshire in 1066. However in 1069 William, first of the Norman kings, found his advance against rebellious northern earls delayed by the bridge over the Aire near Pontefract being broken.

*This medieval bridge at Wareham probably replaced a wooden bridge dating from Saxon times*

*Former Avon bridge at Bristol*

No early timber bridges remain in use anywhere in Britain and their form is known to us only from excavations and old drawings. Parts of a 12th century timber bridge were found under the late 13th century stone bridge at Monmouth. Other Monmouthshire timber bridges at Caerleon, Chepstow and Newport remained in use until replaced in stone in the early 19th century. That at Newport is known from old drawings to have had piers in the form of rows of three posts rising from sills. Diagonal struts seem to have allowed the provision of a deck wide enough for two carts to pass each other. On the Thames the bridges of Windsor and Kingston were of timber throughout the medieval period. An excavation at Hemington in Leicestershire found substantial parts of a late 11th century timber bridge over the River Trent thought to have had a roadway 2.8m wide carried 5.5m above the river on trestles rising from a pair of lozenge shaped caisson bases, ie wooden boxes filled with sandstone rubble. In the early 12th century this bridge was replaced by a bridge with stone footings underpinned by oak piles itself replaced by a mid 13th century structure with four regularly spaced piers across a 50m wide former course of the river.

Clapper bridges consist of large stone slabs either simply spanning a gap between the banks of a small stream or set upon piers composed of smaller slabs carefully laid on top of each other without mortar. These bridges can either be the width of each slab laid in line, or they may have extra width provided by parallel lines of slabs. The slabs are often undressed and can make awkward joints against each other. These bridges usually lie beside fords, so it appears that they were only ever originally intended for pedestrians and perhaps packhorses. Clapper bridges are often described as being prehistoric but this does not mean they are thousands of years old. Historical records and/or provision in the stones for metal ties or hand-rails suggest that some of the clapper bridges in Derbyshire and Lancashire and also in Mayo and Cork in Ireland are 18th and 19th century creations. Some of the numerous clapper bridges of Dartmoor, Exmoor and Bodmin Moor are probably much older but are unlikely to predate the medieval period. There are clapper bridges in North Wales and the Pennines which may be late medieval and Keble's Bridge in the Gloucestershire Cotswolds is probably 16th century. Some clapper bridges on the NW side of Bodmin Moor carry metalled roads and have parapets of concrete or mortared stone, but there is no evidence that any clapper bridges had parapets or handrails of any kind before the late 18th century.

*Keble's Bridge, Eastleach, Gloucestershire*

*Ruined clapper bridge at Dartmeet on Dartmoor*

*London Bridge as depicted in the early 17th century*

Funding bridgeworks has always been difficult. Under the later Saxon kings it was regarded as one of the "trinoda necessitas" or three obligations of freemen, the other obligations being military service and the maintenance of burh fortifications. This remained in force until the time of Magna Carta in 1215, after which only those with an obligation to help maintain a specific bridge were required to do so. In 1285 the Statute of Winchester confirmed that each lord of a manor was responsible for maintaining local sections of the king's highway although bridges weren't specifically mentioned. Kings often granted to a town the right to collect a tax known as pontage for a specified number of years to raise funds for bridge repairs and maintenance. Elsewhere bridge building usually needed at least one major land-holder, lay or ecclesiastical, to act as principal sponsor. Smaller donations and bequests as frequently recorded in the 15th and 16th centuries were only sufficient for minor repairs and embellishments to existing bridges but by the 14th century there were merchants with both the money and potential trade incentives to be major instigators of the construction of new stone bridges.

Bridge building was regarded as an act of piety and it was common for later medieval bishops to grant indulgences to those that contributed towards maintenance costs. So many bridges were built by ecclesiastical authorities that the reformation of the mid 16th century makes a good cut-off point for the period to be considered in this book as medieval from the point of view of bridge construction and maintenance. A bishop of Durham and an abbot of Abingdon were builders of late 12th century bridges at those places from each of which a single round arch is thought to survive. However timber bridges seem to have remained the norm during the 12th century. During this period there were four main roads protected by the King's Peace, three of which were Roman roads still in use: i.e. the Fosse Way, Ermine Street and Watling Street.

One major new stone bridge of the late 12th century was that over the Thames at London begun in 1176 by Henry II to replace an early 11th century timber bridge patched up after storm damage in 1091 and a fire in 1136. Now represented only by a few loose carved stones in St Magnus' churchyard, the new bridge was twenty arches long and 30m upstream of the existing bridge of 1967-72. The difficulty and expense of constructing it would have put the project on a par with major building campaigns at the Tower, on the cathedral, or on the city wall. After the bridge was completed in 1209 King John licensed building plots upon it to help recoup the enormous construction costs. This resulted in half of the eight metres of width being taken up by shops on either side of the two carriageways each shared by pedestrians and vehicles. It was an ideal place to trade as few could miss seeing your wares, but congestion partly caused by loitering shoppers was clearly a serious problem at times. In 1722 a "keep left" rule had to be introduced, which appears to be the earliest record of this system being used on roads in Britain. Despite congestion being an issue, markets were sometimes held on bridges, as at Brigg and Norwich, although a more common arrangement was for a market to develop where roads converged at a bridge-head in or near a town.

There were 138 shops on London bridge by 1358 and eventually timber-framed houses built over the shops (and over the roadway, reducing sections of it to tunnels) rose to another six storeys. Mills using a pair of arches at either end were powered by the tide race since the width of the many piers could at times result in the water level upstream being a metre and a half or more higher than that below. The bridge had gates at either end and a drawbridge in the middle to allow ships to pass. Both private and public latrines overhung the parapets, a new public latrine being built at the north end in 1382-3. Fires damaged buildings on the bridge in 1212, and again during the peasant revolts of 1381 and 1450. After the latter, rebel leader Jack Cade's head was impaled on a spike high up on the south end, part of a custom begun after William Wallace's execution in 1305 and continued until the 18th century, Sir Thomas More in 1535 being another victim. Damage caused by another fire in 1633 left a gap at the north end which saved the bridge from further harm during the great fire of 1666. However two arches were replaced about that time by a single span to improve navigation. The whole structure was dismantled after a new London bridge was built downstream in 1825-31.

Other English towns once had bridges like that at London, although shorter and usually somewhat later in date. Public latrines are mentioned as a feature of the bridge at York, which also had a prison at either end. There navigation was improved as early as 1566 by replacing the central arches with a single 24m wide span. In 1354 there is a mention of granaries on the bridge at Henley-on-Thames. Old illustrations show buildings two storeys high upon each of the bridges at Shrewsbury. At Bristol additional walls 1.2m thick were built on outer arches set on the starlings of the piers. Beams bridged the 5m gaps between the new walls and the 4.5m wide bridge. Upon these beams were built shops which eventually had four upper levels of apartments above them, some with wooden walls cantilevered out from the stone wall below.

Towns took great pride in their major buildings and often depicted them on their official seals. Bridges appear on the medieval seals of the towns of Barnstaple, Bideford, Bridgwater, Cambridge, Colchester, Maidstone, Rochester and Stirling.

Many bridges in towns had gatehouses upon them which formed part of the town defences. The only surviving example is the thin gateway perched on a pier of ordinary size at Monmouth which formed an outwork some distance from the main walled area of the town. The Welsh Bridge at Shrewsbury had a twin towered gateway on the next pier out from the town, whilst a tower at the other end formed an outer gate. A gatehouse also stood until 1702 on the six-arch bridge at Worcester built in 1313-28, the piers of which were only demolished with some difficulty in 1781. A gateway also stood upon the surviving bridge at Norwich. More common was for a gateway to stand on dry land on the town side of a bridge, as with the small square tower surviving at Warkworth in Northumberland. A gateway of this type lay at the west end of the English Bridge at Shrewsbury. As in many town gateways the upper rooms were latterly used as prison cells and in 1546 a prisoner was left dangling from his shackles when flood-waters brought down much of the gate underneath and around him.

*The fortified bridge at Monmouth*

By the 13th century a fashion arose for the building of chantry chapels upon stone bridges. The chapels were seen as emblems of Christian charity, and priests serving them not only ministered to the spiritual needs of weary and wary travellers but also prayed for the souls of major benefactors donating funds towards the costs of bridge-works. Prayers for benefactors were believed to reduce the time spent by the soul in Purgatory before ascending to Heaven. Bridge-chapels still project mid-river from the east sides of bridges at Wakefield, Rotherham and St Ives, the latter two chapels built in the 15th century. At Bradford-upon-Avon a 17th century lock-up has replaced the superstructure of the tiny chapel of St Mary perched above the middle cutwater on the east side. Not much of the bridge itself survives at Derby, where the surviving chapel lay beside the town end of the bridge. At Exeter a church of St Edmund lay beside two arches of the city end of the bridge. It was mostly rebuilt in the 1820s and apart from the tower only ruinous crypts below the nave and aisle remain. The 14th century bridge at Bridgnorth had a chapel at the furthest end from the town. A chapel of St Mary was added in 1361 above the central part of the four-arch bridge of c1240 at Bristol. Set over a room used for council meetings the chapel was 22m long by over 6m wide and had a tall bell-tower raised above the roadway across the bridge. Other 14th century chapels lay on or beside the bridges of Bedford, Huntingdon, Nottingham, Rochester and Stockport. A chapel on the bridge of Bromham or Biddenham in Bedfordshire was founded in 1295. Records suggest that over a hundred medieval bridges in England had chapels either upon them or closely associated with them.

Possibly the earliest of the English bridge chapels was that at London. It was dedicated to St Thomas Becket, the bridge having been begun by Henry II partly in expiation for his part in the murder of the archbishop in Canterbury cathedral in 1170. The single arch High Bridge at Lincoln also had a projecting chapel of St Thomas Becket upon it. Buildings of the 1540s on the other side of the roadway over this bridge still stand several storeys high and are the only survivors of their type in Britain. In Derbyshire the Swarkestone bridge was recorded as having a chapel at the south end in 1558, and ruins of one still stand at the south of Cromford bridge. At Durham a House of Correction of 1632 has replaced a chapel at one end of Elvet bridge. Part of another chapel still remains at the other end, and the Framwellgate bridge also had a chapel set upon it. The early 16th century bridge built by Bishop Dunbar over the River Dee to the SW of Aberdeen also had a gatehouse and chapel of St Mary at the northern end closest to the city. These bridge chapels ceased to be used as religious buildings after the Reformation, all chantries in England being closed as a result of an Act of Parliament of 1547 under Edward VI, and those not demolished saw a variety of later uses.

St Mary's bridge at Tamworth was so named from a figure set upon a cross on one of the parapets, the pedestal of which has been reset on the castle mound next to where the bailey wall crosses its ditch. Plain crosses set on the parapets of pedestrian refuges of bridges were quite common. None survive intact but sockets for mounting crosses can be seen on several bridges, as at Chew Magna, Hampton-in-Arden and Radcot.

The late 13th century bridge at Newcastle bore both a chapel and a prison. Because the River Tyne formed the county boundary the townsfolk took responsibility for the six northern arches and the bishop of Durham maintained the southern four. This was quite a common arrangement in such circumstances and could result in two parts of a bridge being maintained to different standards. At Tenbury Wells on the Shropshire-Worcestershire border the southern arches were rebuilt in the 18th century but the northern arches are still medieval and only in the widening of 1908 was the bridge treated as one unit. Ickford bridge has datestones of 1685 in the middle proclaiming "Here ends the County of Oxon" and "Here Begineth the County of Bucks".

*The ruined Exe bridge at Exeter with the tower of St Edmund's Church*

*The 14th century bridge with a projecting chapel at Wakefield*

*13th century bridge forming part of Fountains Abbey in Yorkshire*

By the 13th century agricultural surpluses were increasing and there was a general increase in trade, especially in wool, textiles and pottery. Water transport was still cheapest, particularly for heavy or bulky commodities, and especially building stone, but inland transport thought to have cost an average of one and a half pence per ton per mile was certainly affordable. This was reflected in the replacement of older timber bridges by new ones of stone. They were better able to carry heavily-laden carts and less likely to be washed away by flood-waters with consequent disruption to trade traffic. A packhorse could carry four bushels (about 150 litres) of grain but a cart might carry up to eight times as much as the payload of one horse.

Heavy usage of carts stimulated the building of stone bridges. In areas where the use of carts was impossible because of a lack of roads there were no stone bridges. There are no medieval bridges at all in the Scottish Highlands and very few in Wales. In Ireland about three dozen medieval bridges survive in English-dominated areas but the few bridges built by Gaelic chiefs all seem to have been of wood or wickerwork.

In some cases only the piers of a bridge were built of stone as had been the Roman common practice. At Mordiford in Herefordshire evidence was found of a former roadway at a level too low for any arches to be formed beneath it. Thus the corbels on the piers close to river level in this instance must have been originally intended to support diagonal struts carrying the permanent roadway rather than temporary shuttering for the construction of stone arches. Wood was not only a lot cheaper than stone to build with because constructing stone or brick arches required a lot of timber for shuttering, but was a lot quicker to put up. The shaping of cut stones and digging out of foundation trenches might take place throughout the year but the building of stone arches using lime mortar could only take place during the few months of the late spring and summer which were likely to be free of frosts.

Unless set on solid bedrock, piers were often supported on staddles or closely grouped piles driven into the river bed. Surrounding wedge-shaped platforms called starlings with cut stone or piles forming their outer edges might be used to protect the piers from the effects of eddies and floating debris. These are rarely now visible above normal river water levels. Massive piers are a feature of some bridges and would be necessary so that each arch was stable in its own right without a counter-thrust being provided by the next arch. Lack of suitable weather or sufficient funds, or even political changes, could mean a long delay before the next arch could be built.

The arches of medieval stone bridges can be round, pointed, or segmental in shape. Arches of a vague shape neither quite truly round nor obviously pointed are quite common in bridges built from rough rubble or with arches formed of thin slabs, perhaps as a result of settlement or hasty later repairs on tight budgets. Pointed arches offered builders a way of reconciling variable arch spans with a need for constant arch height, although some bridges are in any case obviously hump-backed, or have a road rising from one river bank to another, as at Dairsie. The arches of London bridge spanned 6m, a dimension not often greatly exceeded in Southern England, Wales or Ireland.

Over the faster flowing rivers of Northern England with rockier and steeper-sided valleys designers tended to favour a smaller number of wide and high arches needing the minimum number of piers on the river-bed. The 15th century bridge at Twizell in Northumberland has a single arch spanning 30m. Of similar span are the two arches of the Newton Cap bridge near Bishop's Auckland, either late 14th or 16th century, or possibly with an arch of each date. No other bridges in Britain had arches spanning such distances before the 18th century but several bridges in Scotland have arches with spans from 13 to 15m, as at Balgownie, East Linton, Newbattle and North Water. Arches at Kirkby Lonsdale and several bridges in County Durham have similar spans.

*The bridge across the estuary of the River Taw at Barnstaple*

*Newbridge across the River Thames in Oxfordshire*

*The causeway forming the southern approach to the bridge over the River Ouse at Harrold, Bedfordshire*

*Elstead, one of several modest bridges over the River Wey in Surrey*

Some arches only have one arch-ring, but commonly there will be two, and sometimes three. Occasionally the lower arch-ring is recessed, or to be more precise, the upper arch-ring(s) project out over those below. This demanded greater skill in construction and needed the arch-ring stones to be finely-cut ashlar blocks. In such cases the outer edges of such arch-rings may have chamfers. Pairs of arch-rings set in the same vertical plane as each other are common and could be constructed either of ashlar or of wedge-shaped stones called voussoirs requiring varying amounts of dressing by masons depending on the type of stone and how it naturally fractures. See the drawings below.

About two fifths of the medieval bridges still surviving in England and Scotland have ribs formed of carefully cut stones under the soffits of at least some of the arches. The ribs either rise from above an offset in an abutment or pier or die into vertical wall-surfaces. Ribs may start either above or below normal water levels. The outermost ribs normally provide the lowest arch-ring on each face, which may be recessed. Ribs were usually built first and then main arches created over the top of them, sometimes with rougher materials. The ribs often have both outer edges chamfered. In some cases only the outermost edge of the two end ribs facing upstream and downstream have a chamfer. Occasionaly ribs were broken out of the central arches later on to give a few inches more height for vessels to pass underneath, as at Newbridge in Oxfordshire.

Often there will be a string-course between the arches and the parapet at the side of the road. Not all bridges are straight and commonly even a bridge that is itself relatively straight will have a sharp bend on one of the approaches. Parapets may be very low or may never have existed on some bridges, especially small and low bridges on trackways mainly used by pedestrians and pack-horses. Normal waist-height parapets would have caught against the panniers mounted on either side of packhorses, who would have been made to use the bridge rather than an adjacent ford used by vehicles or other animals because pack-animals might have been tempted to sit or roll in the water to cool down, to the detriment of their loads. So-called packhorse bridges continued to be built until the early 19th century. Large numbers of them survive but only a few of them can be regarded as medieval, and most of the surviving examples are 17th or 18th century. Several bridges of this type were actually more to do with giving villagers access to their parish church, as at Bruton, Medbourne and Wilberfoss.

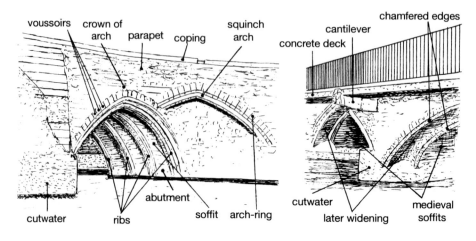

*East Farleigh, Kent*                    *Fordingbridge, Hampshire*

*Aveton Giffard, Devon*

*Wide-ribbed arch at Chester*

*Alresford, Hampshire*

*Stretton, Herefordshire*

*The older arches at Tenbury Wells with modern widening above*

*The bridge over the Great Ouse at St Ives, Cambridgeshire, with its projecting chapel*

Medieval bridge piers commonly have triangular cut-waters to reduce the impact of water flow upon them. On the downstream side they may be round instead of triangular as on the River Wey bridges, or replaced by square buttresses. Cut-waters either die back into the main structure below parapet level or continue up to provide refuges for pedestrians to avoid vehicles. Sometimes triangular cut-waters support semi-hexagonal refuges as at Jedburgh and Kirkby Lonsdale. Commonly a bridge will have refuges on all or most of the cutwaters or none at all, or just two opposed central ones, but more irregular arrangements can occur. Medieval bridges in towns or on main roads could have road-widths before 18th or 19th century widening just sufficient for two carts to pass each other but many country bridges had road widths between parapets of 2.5 to 3.5m, barely enough for a cart and a pedestrian side-by-side.

Quite a number of bridges in the south and middle of England are located in flood plains of rivers where not just a single bridge was needed but a long raised causeway on one or both approaches to it. Examples are Bromham, Harrold, Stafford and Turvey all in Bedfordshire, Mordiford in Herefordshire, Lacock in Wiltshire and the bridges of Stone Canon and Clyst St Mary lying some way out on either side of the city of Exeter. Such causeways have flood-arches and sometimes cut-waters, and can connect a sequence or two or three bridges each composed of several arches over various channels of a river. At Swarkestone in Derbyshire the bridge and causeway together originally had thirty-nine arches, but the five arches which now cross the River Trent itself are an 18th century rebuild. In 1474 Maud Heath of Langley Burrell in Wiltshire left enough property to sponsor the erection and maintenance of a road and causeway together extending several miles from Wick Hill to Chippenham. Sixty-four segmental arches set on thin slab piers now apparently of 19th and 20th century date carry a long raised section of path beside a road now leading to an early 19th century bridge.

Ancient bridges are notoriously difficult to date with precision. Many of them have clearly been much rebuilt or repaired over the years and quite a lot of them have at least one arch that doesn't match the others in either shape or materials or age. Reliable records of major medieval bridge works are infrequent and can be difficult to match up to any surviving parts of a bridge. Neither the shape or width of arches, nor the presence of ribs with or without chamfers, nor the presence of cutwaters with or without refuges is an infallible indication of date. Ribs tend to be regarded as a medieval feature but can in fact be of any date from the 12th century to the 18th century. Consequently several bridges with ribbed arches lie outside the scope of this book.

There is a tendency for 13th century bridges to have arches made of cut stones (as opposed to slabs laid as voussoirs) mainly when associated with towns. Segmental shaped arches seem to be of after 1300 and four-centred arches are likely to be of after 1370. The latter are more likely to be found in towns or in bridges over the moats of castles and houses, as at Eltham and Hedingham. Both these arch forms exerted more horizontal thrust than simple pointed or semi-circular forms so that builders must have had great faith in the strength of their piers. Bridges with triple arch-rings each formed of finely cut stones with chamfered edges are likely to be of c1370-1520, although Edwin Jervoise dated one such bridge at Elvington in East Yorkshire as late as c1700.

Occasionally bridges bear inscriptions or heraldry relating either to original builders or to major benefactors when rebuilding was required. The bridge at Irthlingborough in Northamptonshire has the cross-keys badge of Peterborough Abbey upon it. Panels on bridges are common in Scotland. In Lothian the Pencaitland bridge bears the Sinclair arms, whilst the bridge of the 1530s at Doune bears the arms of Robert Spittal. Panels could easily be reset so are not a safe indication of the wall into which they are set being of the same date as the panel, as at Upper North Water bridge near Montrose.

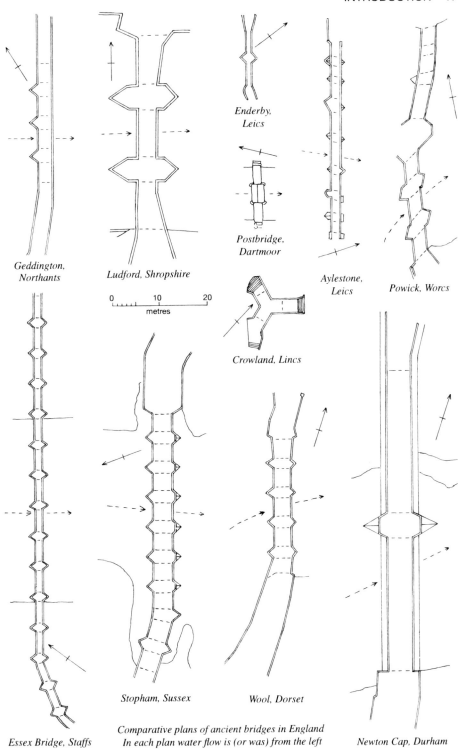

Geddington,
Northants

Ludford, Shropshire

Enderby,
Leics

Postbridge,
Dartmoor

Aylestone,
Leics

Powick, Worcs

0    10    20
metres

Crowland, Lincs

Essex Bridge, Staffs

Stopham, Sussex

Wool, Dorset

Newton Cap, Durham

*Comparative plans of ancient bridges in England
In each plan water flow is (or was) from the left*

In 1531 the Statute of Bridges ordained that in England and Wales where an obligation to maintain a bridge could not be proven then country bridges would be maintained by the shire or riding whilst bridges in towns and cities were the responsibility of those living within the borough boundaries. Justices of the Peace, when sitting in their quarter sessions, were empowered to enquire into unusable bridges on highways "to the damage of the King's liege people", and either force landowners to repair bridges or to raise sufficient funds and labour to effect repairs. This meant that powers were already in place to take over maintenance of many bridges built by monasteries, all of which had been dissolved by the end of 1540. Also, numerous bridges built by bishops were regarded as one-off acts of piety without any inheritable liability for maintenance. The maintenance system established in 1531 remained in force until new county councils were created in 1888 and about the same time many old bridges gained legal protection as scheduled ancient monuments, although this has complicated arrangements that have to be made for their maintenance. Some bridges were not adopted into public ownership and thus liability for maintenance until the early 20th century.

Not all ancient bridges were intended for public use. The abbeys of Bayham, Bury St Edmunds, Fountains and Waltham all have 13th or 14th century bridges remaining within their precincts. Towards the end of the medieval period bridges of stone or brick began to replace timber bridges with a drawbridge or turning bridge at the inner end at many castles and moated houses. The palaces of Eltham and Hampton Court, plus Westhorpe Hall in Suffolk (home of Henry VIII's sister Mary) the royal castles of Leeds (Kent) and Nottingham, and other castles at Hedingham and Pleshey in Essex all have late medieval bridges. Bottom plates for box-frames supporting medieval timber bridges have been found in the excavation of several moated sites and earthwork castles.

Wales and Ireland have a number of bridges of the period 1550-1700 but in England bridgeworks of that period mainly concern repairs to older structures rather than any new projects except for access bridges over moats and drains on private land. Indeed the number of bridges in England used by the general public only began to significantly increase in the 18th century with the creation of new turnpiked roads and canals. No extra bridges were built across either the Severn or the Derwent until that period.

*Bridge at Eltham Palace*

*Bridge at Goodrich Castle*

*Bridge at Leeds Castle, Kent*

*Barnard Castle Bridge, Co Durham*                    *Framwellgate Bridge at Durham*

Quite a number of bridges required rebuilding or repair following damage sustained during the Civil Wars of the 1640s when drawbridges often replaced stone arches. Coach traffic on roads increased during the 18th century and led to the establishment of five hundred new trusts to maintain sections of turnpiked main roads during the period 1750-72. Most medieval country bridges were only wide enough (2.0 - 2.5m) to take one cart or coach. Bridges on main roads soon began to be widened to at least 4.5 between the parapets so that wheeled vehicles could pass each other on them. Modern traffic needs a width of at least 7m to ensure a smooth flow of two-way traffic. In the late 18th century town bridges began to be cleared of gatehouses and shops constricting the roads upon them. By the mid 19th century large numbers of medieval bridges had been either widened or totally replaced, especially those carrying main roads over principal rivers. Just one widened medieval arch at Bridgnorth is all that remains visible of eight bridges each of several stone arches over the River Severn existing by the early 16th century. The Thames and the Great Ouse still each have five bridges with medieval arches surviving, whilst the Warwickshire Avon has six out of fourteen medieval bridges surviving in some part. All four of the medieval bridges over the Tees survive, even if widened and much rebuilt, and five very impressive ones still remain over the River Wear. Of lesser rivers in the south of England, the Rother has seven bridges earlier than 1700, the Wey has five and the Medway four. However survival rates of old bridges in the south-east of England have been poor compared with those of the north.

Signs of joints in the soffits of the arches are generally a clear indication that a bridge seemingly of 19th or 20th century character when seen from above has at least some stonework left underneath the roadway that predates the early 18th century. At many bridges the degree of projection and strength of the cutwaters was deemed sufficient for adding outer arches between them on one or both sides, as at Monmouth, where widening of the bridge by any other means would have destroyed its unique character as a bridge dominated by a medieval gatehouse. The Ludford bridge at Ludlow is a good example of where squinch arches have been employed to widen approaches to assist vehicles in taking tight turns to get onto the bridge. Cantilevering has been employed on some bridges to increase roadway widths or support separate footways (picture, page 14). This tends to destroy their ancient character when seen from the side.

*Trefoiled corbelling on Huntingdon Bridge*          *Paved bed of River Wharfe at Otley in Yorkshire*

Well made stone arches are strong and have coped surprisingly well with heavy loads imposed upon them by today's traffic. In the early 20th century a number of old bridges were discretely strengthened by providing concrete saddles over arches with cracks in them mostly caused by bridge abutments moving slightly away from each other and thus widening spans between them. Occasionally bridges were originally provided with a bed of cobbles or slabs underneath to help prevent scouring away of the river-bed and pier bases, as at Otley and also at the Roman bridge at Piercebridge. A seminar organised in Ireland in 1987 after nine masonry bridges (none of them probably medieval) in Co Wicklow had collapsed after heavy rain the previous year found that all but one had fallen because of scouring of foundations caused by gradual lowering of river-bed levels. The aprons of cobbles or concrete around the bases of piers and abutments of many bridges are an answer to this problem, although they tend to make the bridges less handsome to look at.

Medieval bridges at Cropredy and Chippenham have been lost to rebuilding and widening since the 1930s. A bridge in Tavistock was totally replaced in 1940 to enable tanks access onto Dartmoor for military excercises, and a bridge at Sandon in Staffordshire was lost in 1947. In Bath a ribbed medieval arch from the old bridge was removed during building works in 1964. The only ancient bridges in England totally lost since then are those of Petherton and Woollard in Somerset, the latter a victim of flooding in 1968. The last remains of Bridge of Earn in Scotland survived until the 1970s. These recent losses have been balanced by discoveries of unexpected survivals, such as the bridge found hidden away as part of a long culvert in Rochdale now being brought back into view, and part of a bridge found in a gravel pit at Castle Donnington in 1993.

One problem with old stone bridges is that their parapets may not meet current safety standards requiring firstly sufficient strength to redirect large family cars hitting them at an an angle of up to twenty degrees and secondly they are often barely tall enough to prevent pedestrians or cyclists from accidentally falling over them. Modern general principles of conservation of old bridges recognise that stabilisation should involve the minimum amount of intervention, and when intervention is unavoidable it should be recognisable and reversible at a later date. This generally means that road surfaces need to be kept waterproof and that if possible materials used should be of the same type and in the same manner as when the bridge was first built. It has gradually been recognised that some works done on medieval structures of all kinds in the first half of the 20th century have ultimately done more harm than good.

About a hundred of the bridges described in this book are unwidened structures earlier than 1600 still carrying motorised traffic, in many cases in considerable volumes, but many main roads have been re-routed since the 1960s to bypass old bridges, even ones considerably widened since the medieval period. Generally speaking the most enjoyable medieval bridges to visit are those now only carrying light local traffic or just pedestrians and cyclists. The gazetteers mention where bridges are now bypassed.

*A particularly fine packhorse bridge beside a ford at Sutton in Bedfordshire*

*Bridgend, Glamorgan: a wide single arch replaced two medieval arches in a rebuild of c1780*

*Bridge over the Great Ouse at Barford, of stone, but with unusually ornate 19th century brick widening.*

# MEDIEVAL BRIDGES OF SOUTH-WEST ENGLAND

## AUSTIN'S  SX 750659  East of Buckfastleigh, 6km NW of Totnes

Named after a milling family who built it, perhaps in the 16th century, this bridge has five segmental arches over the River Dart. There is a stringcourse under the parapet.

## AVETON GIFFORD   SX 693470   5km NW of Kingsbridge

The road from Modbury towards Kingsbridge runs over a medieval causeway 360m long. Built c1440, the bridge comprises a middle part with five round arches widened on the upstream side in 1817 (one arch is now blocked up), a northern part of two segmental arches over a mill leat, and a third part to the south. Other arches in the causeway have been blocked up. See photo on page 15.

## BARNSTAPLE   SS 558329   On the SW side of Barnstaple

Thirteen of the sixteen arches over the River Taw are original 13th century work with a road width of under 3m. The three arches at the townward northern end were replaced in 1589. The bridge is approached by causeways, that on the western bank being 450m long. The bridge is mentioned in c1280, 1333, and suffered damage in 1437 and 1646. Cantilevered footways were provided on both sides in 1834 but that on the downstream side was removed in 1963, when a wide new deck was added. See p13.

## BARTON   ST 822605   On the west side of Bradford-on-Avon

This is a packhorse bridge of four pointed arches across the River Avon. The arches each have two arch-rings with the inner one recessed. There are low cutwaters on the upstream NE side and there are iron railings instead of parapets.

## BELLEVER   SX 659772   South of Postbridge, 17km ENE of Tavistock

This was a three-span clapper bridge over the East Dart River, and has been superseded by the modern road bridge just upstream. The central span has long been missing and seems to have once been replaced by planks of wood. There was a similar replacement in wood of the central part of the Fairy Bridge at SX 642725.

## BERRIOW   SX 273757   South of North Hill, 7km SW of Launceston

This medieval bridge over the River Lynher has piers as thick as the three crudely pointed granite arches are wide. The datestone with the year 1640 probably commemorates the provision of new parapets. The bridge was considerably widened in 1890.

## BIDEFORD   SS 455264   On the east side of Bideford

Sir Theobald Grenville is said to have built a bridge here across the River Torridge in the 14th century. The existing structure 200m long of twenty-four arches is thought to be of c1460-1500 but it was extensively repaired in 1638 and widened in 1792-1810 and 1925. An arch at the west end was rebuilt after collapsing in 1968.

## BODITHIEL   SX 176649   8km west of Liskeard

One arch remains of the bridge over the River Fowey seen by Leland in 1538. It has been widened on the west side and two northern arches replaced by one span.

## BOLTER'S   ST 606334   3km west of Castle Cary railway station

A decayed packhorse bridge of three arches carries a track over the River Alham. There are cutwaters rising to full height on the upstream northern side but no parapets.

## BRADFORD ABBAS   ST 589140   4km SE Yeovil

Smith's bridge over the River Yeo has a pair of four-centred arches dating from the 16th century. The low parapet is supplemented by a handrail.

*Bridge over the estuary of the River Torridge at Bideford*

*Berriow Bridge*

*The ruined clapper bridge at Bellever on Dartmoor*

*Bolter's Packhorse Bridge, Wiltshire*

*Barton Bridge over the River Avon near Bradford-on-Avon*

*Packhorse bridge beside a ford at Bury, near Dulverton*

## BRADFORD-ON-AVON   ST 825609   In Bradford-on-Avon, 4km NW of Trowbridge

Most of the existing bridge of nine arches over the River Avon is of 1621, but two 13th century arches remain and a lock-up or short-term prison cell with a domed roof has replaced the chapel of St Nicholas that stood on the central cutwater. In 1757 the celebrated preacher John Wesley spent a night in one of the two cells in the lock-up.

## BRADFORD-ON-TONE   ST 172230   4km SW of Taunton

The bridge with two pointed arches over the River Tone is probably 15th century with repairwork of 1698. The arches have a flush upper ring over two chamfered rings.

## BRIDFORD   SX 805883   10km SW of Exeter

Leland mentions this early 16th century bridge with a pair of four-centered arches over the River Teign. A single wide segmental arch has been built alongside to widen it.

## BUCKFASTLEIGH   SX 744667   South of Buckfast Abbey, 6km NW of Totnes

The old road from Ashburton to Plymouth crosses the River Dart on the much widened Dart Bridge of four pointed arches.

## BURY   SS 911279   East of Dulverton, 15km north of Tiverton

A picturesque four-arch packhorse bridge lies bedside a ford of the River Haddeo. The 2m wide roadway over it is cobbled. The two eastern arches are pointed and the others are segmental. Three cutwaters face upstream but only two face downstream, none of them having pedestrian refuges.

## CARDINHAM   SX 129651 & 107676   6km ESE of Bodmin

The northern of three arches over the River Fowey has a slate arch-ring and may be 15th century, having been seen by Leland. The other arches of granite appear to be a later rebuild. The Lady Vale clapper bridge in Cardinham Woods 3km to the NW has several spans each with four parallel slabs.

*Chew Magna: cross mount   Bradford-on-Avon*        *Widened arch at Clyst St Mary, near Exeter*

**CHEW MAGNA**   ST 575630  South of Chew Magna,  10km S of Bristol

This is a fine late 15th century bridge over the River Tone. Two of the three arches have double arch rings. The other one with a single arch-ring is round and probably more recent. There is a socket for a former cross on the parapet.

**CLAPPER BRIDGE**  SX 351652    NW of Pillaton, 14km NW of Plymouth

Despite its name this late 15th century bridge over the River Lynher has no clappers. Instead it has three semicircular arches of granite, plus a lintelled flood-arch to the SW. Keystones on the downstream SE side suggest considerable later rebuilding. That side has buttresses rather than cutwaters. The central arch on the upstream side has an inner arch-ring with a keystone but the outer arch-ring looks medieval.

**CLYST ST MARY**   SX 972911    4km ESE of Exeter

The 180m long causeway existed by 1238 but the use of Heavitree sandstone in it suggests rebuilding in the mid 14th century. Battles were fought here in 1455 and 1549. On the latter occasion rebels on the causeway with barricades and cannon were attacked from behind by royal forces that managed to cross the river further north by the mill. The eastern river crossing has two segmental arches which have been widened with round arches on both sides. The western river crossing has two arches probably of c1310, each with four ribs and having been also widened on each side.

**CORNFORD**   ST 692120    14km ESE of Yeovil

This 15th century bridge over the Caundle Brook has three pointed arches and a roadway 3m wide with two pedestrian refuges on large cutwaters facing upstream. There are double arch-rings on that side but only single arch-rings on the rebuilt other side.

**CRAWFORD**   ST 919021    1km SE of Spettisbury, 5km NW of Wimborne Minster

This bridge of nine low segmental arches over the River Stour is of c1400 on the north side, where there are four refuges, and of 1819 on the south side, where it has been widened. It has a kink at the south end. There are also three flood-arches of brick. There is a reset datestone of 1719. A bridge here is mentioned in 1235.

**CROSCOMBE**   ST 592442    At Croscombe, 3km WNW of Shepton Mallet.

Two small skewed arches of medieval origin, although widened and rebuilt, cross the River Sheppey. The older downstream eastern side has a buttress.

**CULMSTOCK**   ST 101138    19km east of Tiverton

A bridge here was repaired in 1412. One of the four segmental arches was replaced by two arches in 1780 when two other arches were added. A fifth arch was replaced in the 20th century by a flat lintel. Three refuges face upstream and two face downstream.

**DARTMEET**   SX 672733    25km NE of Plymouth

Until destroyed by the flood of 1826, this bridge had five pairs of clappers across the River Dart. One pier has fallen and just two of the clappers now remain in place. The road now uses a modern bridge downstream to the south. See photo on page 7.

**DRUXTON**   SX 344883    North of Launceston

This 16th century bridge across the River Tamar has four round arches and refuges which must be later since they cut into a stringcourse below the parapet.

## DULVERTON   SX 911279   At Dulverton, 19km south of Minehead

This bridge of five pointed arches over the River Barle is of medieval origin but much rebuilt, having been repaired in 1624, 1866 and 1952. There are small cutwaters. It was widened on the northern upstream side in 1819.

## EASTON GRAY   ST 880873   5km west of Malmesbury

The five low 16th century pointed arches divided by pyramidal-topped cutwaters crossing the River Avon were doubled in width in the 18th century.

## ELLERHAYES   SS 976012   10km NNE of Exeter

This 15th century bridge over the River Colm has four pointed arches which have been widened on the upstream SW side and two flood-arches, one of them of brick.

## EXETER   SX 906922   At the SW end of Exeter city centre

The bridge of c1196-1214 now lies isolated and dry in the middle of a traffic roundabout, its remains being exposed after wartime bomb damage. Originally it was 180m long with seventeen arches. Some of the arches are round and have wide plain ribs. Others are slightly pointed and have polychromatic ribs with chamfers. On the northern side the cutwaters are assymetrical, being angled towards the former current flow from the NW. Beside the two eastern arches lay the chapel of St Edmund of which the undercroft remains, and part of the SW tower built in 1448-9. The rest of the chapel was mostly rebuilt in 1834 and was gutted by fire in 1969. Originally there was a chantry chapel at the other end of the bridge. A new bridge built in 1770 nearby only lasted for five years. The Roman river-crossing lay further upstream. See p11.

## FORDTON   SX 840990   1km south of Crediton, 10km NW of Exeter

Four pointed arches over the River Yeo were mentioned in 1668 and widened c1910.

## FRESHFORD   ST 791599   4km west of Bradford-on-Avon

Leland c1540 referred to "three faire new arches of stone" over the River Frome here. They survive, although the presence of keystones suggests some later rebuilding.

## GALLOX   SS 990432   To south of Dunster, 3km SE of Minehead

A short and low causeway beside a ford on the River Aville contains two low and unequal segmental arches probably dating from the 14th century

## GARA   SX 724564   SW of Diptford, 9km SW of Totnes

The pointed arch over the River Avon has been widened on the downstream side. The north side has two arch-rings with the inner one recessed and outer one chamfered.

*Greystone Bridge over the River Tamar on the Cornwall-Devon boundary*

*A polychromatic arch at Exeter*                    *Helland Bridge, Devon*

## GREYSTONE   SX 369804   6km SE of Launceston

Tavistock Abbey contributed towards the construction of this 67m long bridge of 1439 of four round arches with pedestrian refuges between them and a flood-arch at either end. The River Tamar here forms the boundary between Devon and Cornwall. The 3m wide roadway was paved. Each arch has two arch-rings with the inner one recessed.

## GUNNISLAKE   SX 433722   NE of Gunnislake, 5km WSW of Tavistock

New Bridge has seven pointed arches of granite built by Sir Piers Edgcombe c1520 crossing the River Tamar, here lying in a deep valley and forming the boundary between Devon and Cornwall. The bridge is 55m long and rises 12.5m above the river-bed. There are two arch-rings to each arch with the inner one recessed and the outer one chamfered. Here Sir Richard Grenville's Royalist force was defeated in July 1644.

## HARFORD   SX 506768   3km NE of Tavistock

Much of this bridge over the River Tavy was rebuilt in 1668 and the round central arch may be of that period. The sharply pointed outer arches must be medieval in origin. There are refuges on the downstream south side. The north side has been widened.

## HARTLAND   SS 246248   West of Hartland, 20km west of Bideford

Probably built by Hartland Abbey, and mentioned in 1605, this is a two-arch bridge over the Abbey River. It has been widened on the east side, where there are putlog holes for centering. Abbey bridge further west seems to be 17th century in its present state.

## HASELBURY   ST 459110   11km SW of Yeovil

A well-preserved 15th century bridge crosses the River Parrett with two arches each with chamfered arch-rings and four chamfered ribs. The busy A30 now bypasses it.

## HELE   SS 933278   2km east of Dulverton, 19km south of Minehead

This 16th century bridge of three four-centred arches across the River Exe was repaired in 1628 and 1866, and widened on the downstream side in 1892.

## HELE   ST 186246   3km west of Taunton

The three 15th century arches over the River Tone have recessed lower arch-rings. The bridge was widened in the early 20th century and has parapets of 2002.

## HELLAND   SX 065715   4km north of Bodmin

This bridge with cutwaters and pedestrian refuges over the River Camel has two pointed western arches which may be 14th century. The round eastern two arches are 16th or 17th century, which is also perhaps the date of the corbelling under the parapets over the western arches. There are double arch-rings of thin slate voussoirs.

## HOLNE    SX 730706 & 711709    On SE side of Dartmoor, 12km NW of Totnes

The Old Bridge of c1413 has one segmental arch and three others which are semi-circular. It has been widened slightly. The New Bridge to the west of the same period, also over the River Dart, has two large semicircular arches plus a third smaller one.

## HORRABRIDGE    SX 512700    12km SSE of Tavistock

This well preserved 14th century bridge over the River Walkham has three pointed arches each with two arch-rings of thin slabs and a refuge set upon a cutwater.

## HORSEBRIDGE    SX 400749    8km west of Tavistock

A roadway 3.6m wide goes over a 60m long bridge of seven semicircular arches across the River Tamar dating from 1437. The arches have double arch-rings. The cutwaters on the upstream northern side are longer than those on the downstream side. The bridge is very similar to the contemporary bridge at Greystone further upstream.

## HUCKWORTHY    SX 591505    11km SE of Tavistock

This late 15th or early 16th century bridge over the River Walkham has three arches. On the eastern upstream side part of the parapet is corbelled out at the south end. The northern arch is narrower than the other two.

## IFORD    ST 801588    2km SW of Bradford on Avon

Opposite Iford Manor is a two-arch bridge of c1400 over the River Frome. Upon one of the pedestrian refuges is mounted a modern effigy of Britannia.

## ILCHESTER    ST 523228    7.5km NE of Yeovil

Ilchester was once a place of greater importance than now, with defences and several churches. It has one of England's earliest stone bridges, thought to be of c1200. The seven arches (including one over a millstream) have been much rebuilt and the bridge has lost its medieval character as a result of 19th and 20th century widening.

## JULIAN'S    SZ 004998    At the west end of Wimborne Minster, north of Poole

The eight pointed arches across the River Allen each with ribs are late 15th century. There are three refuges on each side. The bridge was much repaired in the 1630s and was widened on the downstream southern side in 1844.

## KENTSFORD    ST 058427    1km SW of Watchet

A packhorse bridge of two very low arches takes a track over the Washford River. A reset stone on the south side bears a cross carved upon it.

## LACOCK    ST 923681    5km south of Chippenham

This bridge was originally built by the nearby abbey, a house of Augustinian nuns, but was much rebuilt after being too ruinous for carts to use for much of the 17th century. There are four pointed arches over the River Avon and five arches over a tributary to the east with a causeway between them which is solid except for square flood openings. Two arches in the eastern section are a brick-lined rebuild of 1809. The western part has a square buttress facing downstream and both parts have cutwaters facing upstream to the north. The pilaster buttresses over the cutwaters are Georgian.

## LANDACRE    SS 816361    13km SW of Minehead

This late medieval bridge of five pointed arches each with two arch-rings over the River Barle was restored in 1875. Small starlings support tiny cutwaters.

*Harford Bridge, near Tavistock, has a rebuilt central arch*

*Huckworthy Bridge, near Tavistock*

*Lacock Bridge, Wiltshire*

*Horrabridge, near Tavistock*

*Load Bridge*

## LAUNCESTON   SX 328851   1km NE of Launceston

North of St Thomas's church is the West Bridge over the River Kensey. It is a packhorse bridge with refuges between two slightly pointed arches and a third arch to the south.

## LERRYN   SX 138567 & 141571   4.5km SE of Lostwithiel

There are two bridges here both dating from the 16th century. One is a single arch of slabs. The other, over the River Lerryn, was rebuilt in 1575 but is older in origin, having been mentioned c1540 by John Leland.There are two unequal arches. The coping of the western pedestrian refuge bears an inscribed date 16--. See page 32

## LOAD   ST 467238   13km NW of Yeovil

The downstream face has been rebuilt of this 15th century bridge of five pointed arches across the River Yeo.  Upstream each arch has a chamfered arch-ring and a chamfered rib. The central arch is higher and perhaps later.  The railings date from 1814.

## LOSTWITHIEL   SX 406598   To the east of Lostwithiel

The River Fowey is crossed by an early 14th century bridge of four pointed arches with double arch-rings and a slightly later four-centered arch to the west of them. There are 18th century flood arches at the east end. The parapet dates from 1676.

## LYDFORD   SX 509846   10km NNE of Tavistock

A single rubble pointed arch, possibly of the 13th century, although not mentioned until 1478, spans a deep narrow gorge of the River Lyd below the former town. 20th century widening on the north side is with dressed granite.

## MURTRY   ST 764499   2km NNW of Frome

This bridge over the Mells River has been much widened but two pointed and ribbed 15th century arches survive on the upstream western side.

## NEW BRIDGE   SX 349867   3km NE of Launceston

In 1504 Bishop Oldham granted an indulgence of 40 days to contributors towards this four-arch granite bridge over the River Tamar. The three largest arches each span 8m. Recesses above the pier imposts may have held hatch or sluice gates.

## NEWBRIDGE   SX 348680   2km SW of Callington, 16km south of Launceston

The late 15th century bridge with five semicircular granite arches over the River Lynher was widened in 1874 and partly rebuilt in 1898. There is a thin band over the two arch-rings. The eastern arch is smaller than the others. The western pier has no cutwaters.

*The bridge over the River Fowey at Lostwithiel. The right-hand arch is a later rebuild.*

### PANTERSBRIDGE    SX 159680    7km ENE Bodmin, west of St Neot

This 15th century granite bridge over the River Bedalder has two pointed arches with two arch-rings of thin slabs. The roadway is 3m wide and has a refuge on each side.

### PENSFORD    ST 619638    10km SSE of Bristol

The bridge over the River Chew was much rebuilt after the devastating flood of 1968 which destroyed a bridge at Woollard 1.5km downstream to the NE. The north arch is round and cuts into the band of brown stone below the modern parapet. Medieval are the pointed middle arch and the more sharply pointed southern arch. Nearby is a 16th century house built over an arch over Salter's Brook where it joins the river. An arch of the widened bridge at Publow 500m downstream also appears to be medieval.

### PLUSHABRIDGE    SX 303724    Near Pensilva, south of Launceston

The three 15th century arches across the River Lynher were widened in 1913.

### PLYM    SX 523587    5km NE of Plymouth

The five existing segmental arches over the River Plym are 18th century but one pier has traces of springing of one of the medieval arches. This bridge once had a chapel.

### POSTBRIDGE    SX 645789    At Postbridge, 20km ENE of Tavistock.

A fine clapper bridge of three spans crosses the River Dart just downstream to the south of the modern road bridge. See page 17.

*New Bridge over the River Tamar near Launceston*          *The bridge over the Lydford Gorge*

*Respryn Bridge, showing the squinch arches allowing partial widening.*

*Lerryn Bridge: abutment*

## RAMPISHAM    ST 561023    12km NE of Bridport, 20km NW of Weymouth

This 16th century bridge over the River Frome has a pointed arch to the north of a pair of four-centred arches with buttresses on the upstream side.

## RESPRYN    SX 099635    4km SE of Bodmin

Of five arches over the River Fowey the pointed middle one may be 15th century, whilst those to the east, one pointed and the other round, are probably 16th century. The western arches are 19th century, as are the square refuges on the downstream south side and the various squinch arches allowing partial widening. This bridge is mentioned as early as c1300, when it had a chapel of St Martin on or near to it.

## ROTHERN    SS 479198    1.5km NW of Great Torrington, 7km SSE of Bideford

The Rolle bridge of 1928 enables A386 to bypass this 15th century bridge of four slightly pointed arches over the River Torridge. It has been widened with round arches out onto the cutwaters on both sides.

## RUTHERN    SX 013669    6km west of Bodmin

This must the "Rothyn Briygge" over a tributary of the  River Camel mentioned in 1494. There are two arches with double arch-rings of thin slate slabs and a cutwater with a pedestrian refuge just on the upstream side.

## ST AUSTELL    SX 009522    SW of the centre of St Austell

This 16th century bridge has three round arches over the St Austell River and pairs of tiny refuges. The middle arch has a recessed inner arch-ring.

## ST ERTH    SW 549 350    9km NE of Penzance

John Leland in 1538 referred to this bridge over the River Hayle as being two hundred years old. It has four low semicircular arches and was heavily restored in the 17th century and widened in 1816.

## SIDFORD    SY 138900    3km ENE of Sidmouth

This medieval bridge of two segmental arches has been widened with concrete in 1930 but with the original parapet on the widened side left in situ.

*Sturminster Newton*

*Treverbyn Bridge*

*The widened Rothern Bridge over the River Torridge*

## STANTON DREW      ST 596634 & 596636      9km south of Bristol

The bridge over the River Chew was damaged by the floods of 1968. It has two arches with ribs with big chamfers and a worn panel on the east side. A ford at Upper Stanton Drew 1.1km ESE has a raised pavement beside it which contains a low pointed medieval arch which is dry most of the time as the stream now runs under the road.

## STOFORD      ST 568135      3km south of Yeovil

The existing two slightly pointed arches across the River Yeo are probably those seen by John Leland in the 1530s. There is a handrail in addition to a low parapet.

## STAVERTON      SX 784637      6km NW of Totnes

This is a fine seven arch bridge of 1413 over the River Dart but unfortunately there is no access to the sides. The refuges include a rectangular one at the NW end.

## STAVERTON      ST 855610      3km east of Bradford-upon-Avon

Leland mentions this four-arch bridge over the River Avon. It was rebuilt and widened in the 18th century. One medieval arch has a recessed lower arch-ring.

## STOKE CANON      SX 938975      5km ENE of Exeter

The 240m long causeway is of late 13th century origin but most of the nine arches recorded in the 17th century were rebuilt c1810. Two 15th century segmental arches still span the River Culme not far above its confluence with the River Exe.

## STURMINSTER NEWTON      ST 784136      13km SW of Shaftesbury

On either side of this late medieval bridge of six arches over the River Stour 17th century widening has resulted in new refuges being corbelled out from the older piers.

## STURT   ST 731881      SE of Wickwar, 20km NE of Bristol

This is a 15th or 16th century packhorse bridge of two arches over the Little Avon with the western one slightly pointed. There are tiny cutwaters.

## TADDIPORT   SS 488187   SW of Great Torrington, 8km SE of Bideford

A medieval bridge of three arches with cutwaters over the River Torridge has been widened with extended piers on the downstream western side. The northern arch is wider than the other two. The parapet with stones set vertically must be later. Close to the south end lies the chapel of the leper hospital of St Mary Magdalene.

## TARR STEPS   SS 867321   17km SW of Minehead

This is a fine clapper bridge 54m long with seventeen low double spans of clappers crossing the River Barle beside a ford. Leaning stones help to displace the force of the river on the drystone piers. The bridge was mostly rebuilt in its former form by Somerset County Council after the floods of 1952 displaced all the clappers except one.

## TIDEFORD   SX 347595   8km west of Saltash

The Old Bridge over the River Tiddy is mostly 18th century but retains one 15th century arch with chamfered ribs under the outer edges.

## TREKELLAND   SX 301799   10km SW of Launceston

A pair of four-centred granite arches carry a road 3m wide over the River Inny. They may date from 1504 when Bishop Oldham granted an indulgence to contributors. There is a refuge on either side. A flood arch on one side is narrower than the other arches.

## TREKENNER   SX 339770   8km south of Launceston

This 16th century packhorse bridge over the River Inney has three segmental arches and cutwaters.

## TREVERBYN   SX 206675   6km NW of Liskeard

Indulgences granted by Bishop Stafford of Exeter in 1412 probably paid for work on the two slightly pointed eastern arches of this bridge over the River Fowey. The eastern arch has been slightly widened. A third arch at the west end has been widened on both sides and lies beyond a very large cutwater on the upstream northern side. See p33.

## TRURO   SW 827449   East of Truro Cathedral

A medieval bridge of two pointed arches on rubble plinths over the Truro River lies hidden away by later widening both sides. A bridge is recorded here in the 13th century.

## WADEBRIDGE   SW 991724   On the NE side of Wadebridge.

This 96m long bridge of fifteen segmental arches over the River Camel was once two arches longer. It is said to have been built c1468 with John Lovebond, Vicar of Egloshayle as chief patron and instigator. It has been much rebuilt and in 1847 was widened out onto the cutwaters to give a road width of 4.5m, but a further widening with a new concrete deck on cantilevers was needed in 1962-3. King's Chapel lay at the east end and St Michael's chapel lay at the western end.

## WALFORD   SU 011002   Just north of Wimborne Minster

The seven-arch bridge is 16th century in origin, although keystones to the arches on the west suggest later remodelling. The east side was widened in the 18th century.

## WAREHAM   SY 921879   North side of Wareham, 9km SW of Poole

The north bridge has one pointed arch and two other arches, one of which has plain ribs. The parapet is modern and the bridge has been widened. The seven-arch south bridge was replaced in 1924, when a 10th century sword was found in the riverbed.

*The Tarr Steps clapper bridge on Exmoor*

*Widened medieval arches supporting a modern concrete deck at Wadebridge in Cornwall*

*White Mill Bridge in Dorset*

## WELLOW   ST 741581   6km south of Bath

Although much restored, this is a late medieval packhorse bridge across the Wellow Brook with two round arches with a buttress between them facing east and a cutwater facing west.

## WHITE MILL or STURMINSTER MARSHALL   SZ 958005   11km NW of Poole

This fine late medieval bridge over the River Allen has eight round arches each with four chamfered polychromatic ribs. The central arches are higher than the others. The parapets are carried on corbels and have pedestrian refuges on the cutwaters.

## WOOL   SY 844872   7km west of Wareham

This 16th century bridge has four arches each with three plain ribs over the River Frome and an early 19th century flood arch at the south end. There are cutwaters with pedestrian refuges on both sides but those on the eastern downstream side are quite small.

## WYKE   ST 656340   2km NE of Castle Cary, 20km NNE of Yeovil

Two small pointed arches, each with two flush arch-rings, cross the River Brue, with a cutwater between them. The roadway 3.6m wide has posts and rails, not parapets.

## YEOLMBRIDGE   SX 318873   3km north of Launceston

Possibly the oldest surviving Cornish bridge, this 14th century structure has two pointed arches across the River Attery, each with three chamfered ribs, the only ones of their type in Cornwall. The roadway has been widened. Flood arches to the south were rebuilt in the 19th and 20th centuries.

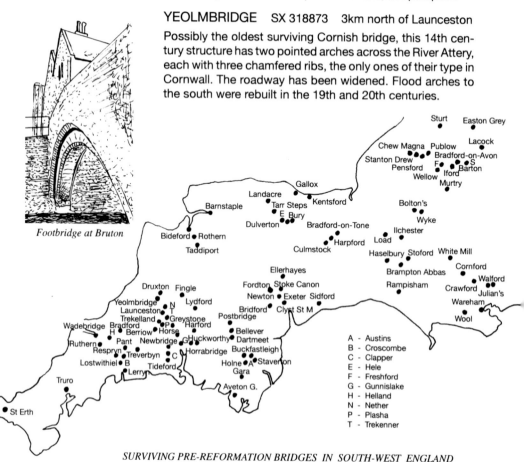

*Footbridge at Bruton*

A - Austins
B - Croscombe
C - Clapper
E - Hele
F - Freshford
G - Gunnislake
H - Helland
N - Nether
P - Plasha
T - Trekenner

*SURVIVING PRE-REFORMATION BRIDGES IN SOUTH-WEST ENGLAND*

*Wyke Bridge*

*Wool Bridge: finely moulded parapet coving*

## OTHER BRIDGES POSSIBLY EARLIER THAN c1600 (not on map).

ALLERFORD  SS 905469    Two arch bridge 6km west of Minehead.

ALTARNUN  SX 223813  Two round arches over Pentpont Water.

BRADFORD   SX 119755   Two clapper bridges, one of four spans carrying a road.

BRENT MILL  SX 696596 S of South Brent.  Widened packhorse bridge.

BRIGHTLY SX 599974  Three four-centred arches, much rebuilt, widened. R.Okement.

BRUTON  ST 684347 Single arch in the form of one wide rib spanning the River Brue.

CHAGFORD   SX 694897   Three arches of granite ashlar of c1600 over R. Teign.

COMPTON DANDO  ST 646646   Five round arches with cutwaters over River Tone.

CORFE CASTLE  SY 957823  Four-arch bridge over ditch in front of castle gateway.

DELFORD  SX 114759   Clapper bridge carrying metalled road over De Lank River.

DENHAM   SX 478679   Single pointed arch of uncertain date high above gorge.

DOWLISH WAKE  ST 376127  Two round arches with low cutwater to east.

FENWORTHY  SX 666841  Single-span clapper bridge usually under waters of lake.

FIFEHEAD NEVILLE  ST 772111  Two pointed arches with cutwater facing upstream.

FINGLE  SX 742899  Two segmental arches across the R. Teign may be of c1600.

FOXWORTHY  SX 757820  Clapper bridge with two spans each of two parallel slabs.

HARPFORD  ST 107214  Probably medieval, widened both sides. Segmental arches.

HOLME  SY 890866   Four segmental arches, one rebuilt. Scene of a battle in 1644.

HUNTINGDON  SX 657622  Clapper bridge of two spans over River Avon.

IVYBRIDGE    SX 636564  Widened slightly pointed arch over River Erme.

LYDIA    SX 696607    Widened single arch, mentioned as decayed in 1669.

MALMSMEAD  SS 791477   Two round arches with cutwaters between them.

STURMINSTER MARSHALL  ST 946001  17th century cambered arch without parapet.

SWEETHAM  SX 881989  Two pointed arches. One arch-ring with band over.

TEIGN-E-VER  SX 661877    Clapper bridge of two spans.

TEIGNHEAD  SX 639844   Clapper bridge of three spans, each two slabs wide.

TOPSHAM  SX 733511  One pointed arch is widened and older than the rest. R.Avon.

TOTNES  SX 806604  Cutwater bases upstream of the 1826 bridge visible at low tide.

WALLABROOK  SX 654871   Single-slab clapper bridge over the Walla Brook.

WENFORD  SX 085751  Two round arches across the River Camel.

WEST LUCCOMBE  SS 899460   One wide elliptical arch, possibly late medieval.

WINSFORD  SS 905349  Medieval arch over Winn Brook much restored in 1952.

The towns of Bath, Bridgwater, Bristol, Chippenham, Looe, Tavistock and Tiverton all once had notable medieval bridges.

# MEDIEVAL BRIDGES OF SOUTH-EAST ENGLAND

**ABINGDON**   SU 497968  &  499969   On the south side of Abingdon

Ock Bridge is mentioned in 1101 when Abbot Faritious took off his shoes here to walk barefoot into his abbey. It was then of wood but one of the surviving round arches may be 12th century. Five of the arches are late medieval and are four-centered with chamfered ribs. The arches are widely spaced and were widened in the 18th century and again in 1979-80. They can only be seen properly from the water. A brick cottage beside the bridge may stand on the site of a hospital chapel of St Mary Magdalene mentioned in the 14th century.

Burford bridge was built in 1416-22 by local merchants to attract traffic over the Thames into Abingdon. Three extra arches to the east known as Maud Hales' Bridge were added in 1430. The complete scheme comprised the two parts of this bridge, that at Culham and a causeway over Andersey Isle linking the two. Untill a rebuilding in the 1920s in which some old parts were retained or reset, Burford bridge contained four medieval arches, two others having been replaced by a wider arch in 1790. There is another Burford bridge in the Cotswold village of Burford, on which see below.

**ALRESFORD**  SU 588331   11km ENE of Winchester

Hidden from public view and only visible from the garden of a nearby house is a lofty single pointed arch over one of the headwaters of the River Itchen. Although said to be part of Bishop Lucy's works here in the 1190s, concave chamfers on the arch make a 14th century date much more likely. See picture on page 15.

**AYLESFORD**  TQ 729589   4km NW of Maidstone

The bridge over the River Medway is 14th century. The wide central arch is of 1824 and has replaced two of the original seven arches. Buttresses support pedestrian refuges on either side of a road 3.6m wide.

**BURFORD**   SP 253125   At the NW corner of Burford, 28km WNW of Oxford

Pontage was granted for the repair of a bridge here in 1322. It was subsequently re-built with four very low segmental arches with refuges between them. The bridge was altered in the 18th century and the parapets rebuilt c1945. From the south the second arch has chamfered ribs and the first and fourth arches have ribs without chamfers. This bridge is not to be confused with another Burford Bridge at Abingdon.

**CHERTSEY**   TQ 051671   NE of Chertsey, 3km NW of Weybridge

The Abbey Mill bridge of three arches over the Abbey River is medieval but has been widened on the SE side using some brick. The parapets are probably of the 1920s.

**CHIDDINGFOLD**   SU 983357   2km east of Chiddingfold, 10km NE of Haslemere

This is a small two-arch 15th century bridge of stone and brick. The downstream north-ern side has been concreted.

**CHISELHAMPTON**   SU 574988   10km SE of Oxford

This 53m long bridge of eight arches over the River Thame is mostly post medieval. Pontage was granted for repairs in 1444. Leland described it a century later as having a timber superstructure on large stone cutwaters, five of which still remain. It was bro-ken down shortly after Prince Rupert fortified it and returned over it after his victory at Chalgrove. The bridge was repaired in 1690 and widened in 1899.

*Aylesford, Kent: a wide modern arch has replaced the original central arches*

*Burford Bridge in the Cotswolds*

*The bridge over the Avon at Christchurch in Dorset*

*Bridge over the River Wey at Eashing in Surrey*

## CHRISTCHURCH   SZ 161922 & 160924 & 137935   5km east of Bournemouth

The bridge over the River Avon east of the castle was the site of a skirmish in 1645. It firstly has two renewed arches, then there is a separate section with five stone arches with cutwaters and double arch-rings (see page 40). It was repaired in 1598 and widened on the upstream northern side in 1900. Finally there is the Waterloo Bridge of five arches dating from 1816. Place Mills bridge south of the priory has two round arches with a head between them. To the NW of the town the 17th century Iford bridge over the River Stour has four stone arches in the centre and others of c1780-1800 of brick.

## CLATTERN   TQ 179691   West side of Kingston-upon-Thames

Three slightly stilted round arches each with two arch-rings over the Hogsmill River may be as old as the 12th century. The upper parts are 18th century brickwork and the cast-iron railings are mid 19th century, when the bridge was widened on the south side.

## COMBE BISSETT   SU 109263   5km SW of Salisbury

The 16th century bridge over the River Ebble has three pointed arches, wooden parapets, and original mounting blocks at either end. It was widened in the 19th century.

## CULHAM   SU 501959   1km south of Abingdon

This bridge of five arches over the River Thames formed part of an early 15th century scheme by Abingdon merchants to bring trade into that town (see above) . The bridge was widened in the 18th century and is now bypassed by a bridge of 1927-8.

## DURFORD   SU 782233   4km east of Petersfield

This bridge over the River Rother has four round arches each with three plain ribs. Chamfering only occurs on the upstream western side of the second arch from the south and the rib next to it. The other arches may be a 17th century rebuilding, which could also be the date of the parapets. There may have been slight widening on the downstream side.

## EASHING  SU 946438   2km west of Godalming

The bridge over the River Wey has segmental arches and cutwaters pointed on the SW upstream side and rounded on the downstream side. From the east there are three arches, then a central island and then another four arches with 18th century brickwork above them.

**EAST FARLEIGH    TQ 734535    3km SW of Maidstone**

This 14th century bridge over the River Medway of five arches of ragstone was regarded by Jervoise as the finest medieval bridge in southern England. The northern arch has a chamfer only on the east side. The other arches have chamfered ribs. The second arch from the south is lower than the others. There is a squinch arch on the west side at the southern approach (see page14). The bridge has a slight bend in the middle.

**EASTLEACH    SU 201052    18km NW of Cirencester**

Keble's Bridge is a five-span clapper-bridge with steps from a pier down into the River Leach. Named after a local family, it may date from the 16th century. See page 7.

**ELSTEAD    SU 904438    6km west of Godalming**

The bridge over the River Wey has five arches and round cutwaters on the downstream side. There is a modern bridge close by to the north carrying eastbound traffic. The arches vary in composition, using differing combinations of thin wedge-shaped stones, cut blocks and larger blocks. There are later parapets of brick. See photo on page 13.

**FITTLEWORTH    TQ 009182    12km SW of Horsham**

The four-centered side-arches of the bridge over the River Rother are 16th century. The higher round centre arch is of 1791, when the bridge was widened on the downstream east side, where the cutwaters may be the medieval ones reset. The bridge over the mill-stream to the north is 19th century.

**FORDINGBRIDGE    SU 149142    At the south end of Fordingbridge**

The medieval bridge of seven flattish pointed arches with chamfers over the River Avon was widened in 1841 out onto the cutwaters of both sides. The arches are polychromatic and may go back to the 13th century. A pedestrian walkway is now cantilevered out from the upstream NE side.  See illustration on page 14.

**FRENSHAM MILL    SU 849420    4km south of Farnham**

The five widened arches over the River Wey were rebuilt using the old materials in 1905. The arches are of brick and have rectangular buttresses oln the downstream side.

**GODSTOW    SP 484092  &  481091    4km NW of Oxford**

One arch of the Old Bridge by the Trout Inn may be medieval. The University authorities met Queen Elizabeth here when she visited Oxford in 1566.  The small pointed arch over a stream west of the Abbey is also medieval.

**GOMSHALL    TQ 084479    East of Guildford**

This is a three-arch packhorse bridge over the River Tillingbourne. The central arch is the largest and has rough block voussoirs.

**HABIN    SU 809229    6km east of Petersfield**

The southern two of the four round arches over the River Rother have widely spaced chamfered ribs  and may be 16th century. The other arches are 18th century. There are small cutwaters on the upstream western side and rectangular buttresses on the east side. The parapets appear to have original re-used coping stones.

*Buttresses at Habin in Sussex*

## HADMANS   TQ 865424      13km west of Ashford

Two 15th century arches each with two ribs and two arch-rings span the River Beult. The bridge is now bypassed.

## HANWELL   TQ 151801     West of Ealing, 15km west of central London

Just two of the six arches over the River Brent are now medieval. The other arches were rebuilt in the 18th century and widened on the north side in the 19th century.

## HAWKENBURY   TQ 799445   1km  NE of Staplehurst, 12km south of Maidstone.

From the north the bridge over the River Beult has firstly a low round arch, then a four-centred late medieval arch and then two round 18th century arches.

## HENLEY-ON-THAMES   SU 763826     At the south end of the town

The Angel Hotel stands on a buried former flood arch of the medieval bridge destroyed by flooding in 1774. Another arch is also reported to survive on the east bank.

## HERSTFIELD   TQ 782468   10km south of Maidstone

This 15th century bridge has two brick arches over the River Beult and three more in a causeway. The parapets are corbelled and of stone.

## ICKFORD   SP 649065   13km east of Oxford

A bridge here is mentioned in 1368 but the two main arches, one four-centered and the other elliptical, are later.  Another arch further west is 18th century. The north side has a buttress and a refuge within which are faint date stones of 1685 proclaiming  "Here ends the county of Oxon" and *Here Begineth the County of Bucks"

## IPING   SU 854229   3.5km ENE of Midhurst

This is a curved five arch bridge over the River Rother. The southern part has small cutwaters with stepped tops on each side. The northern part has one large cutwater on the west side and two rectangular buttresses on the east side. The last arch at that end has a single rib on each side.

## LADDINGFORD   TQ 689498     9km SW of Maidstone

The upper parts have been rebuilt in brick but the two arches over the River Teise are medieval. They are just 2.5m wide.

*Radcot Bridge over the River Thames in Oxfordshire. Note cross-base over central arch*

## LOWER HEYFORD   SP 479248    10km WNW of Bicester

Frst recorded in 1255, the bridge is shown on an old map dated 1606 as having four arches over the River Cherwell and four more in a causeway to the east. Much of the widened existing structure is 19th century brickwork, but two late medieval arches divided by buttresses survive in the main bridge, and others in the causeway.

## MAIDSTONE   TQ 761554    On SW side of Maidstone

Half-hidden within culverting between the parish church and the archbishop's palace is part of a 14th century bridge over a tributary of the River Medway.

## NEWBRIDGE   SP 403014    11km WSW of Oxford

Standing at the junction of the River Windrush with the River Thames, this is the finest medieval bridge in Oxfordshire. It still has six "great arches of stone" mentioned by Leland, with cutwaters with refuges on the upstream side. The second and fifth arches each have two widely spaced ribs and were part of John Galafre's rebuilding in the 15th century. In 1644 Lord Essex was unable to take the bridge in May but in June Thomas Waller's forces overpowered the two hundred Royalist musketeers guarding it. The bridge was broken by Parliamentary forces later that same year and the two middle arches must be of after then. One is lower than the other arches and the other higher than the rest to allow navigation under it. South of the bridge is an island bearing a inn and then beyond that a low causeway in which are six flood arches. See page 13.

## OXFORD   SP 514055, 521061, 503062, 501062, 508063   Around Oxford

A three-arch bridge of 1825 over the Thames has replaced the Grandpont or Folly Bridge, which formed part of a 700m long causeway first built c900, rebuilt in the late 11th century by Robert d'Oilly,  and which had over forty flood arches by the 16th century. One buried arch is said to remain. The place of one arch was taken by a drawbridge. There was a similar feature at the east end of the Magdalen Bridge over the River Cherwell which had twenty stone arches and boldly projecting cutwaters. It was replaced in 1772-8. The Osney Bridge over the Thames is now an iron structure of 1889, replacing three stone arches later widened. Further out are the Seven Arches, round arches probably of 16th century date later considerably widened on the south side.  An iron bridge of 1861 has replaced the Hythe Bridge originally built of wood in the 13th century by Osney Abbey, and replaced by a stone structure c1373-1403.

## RADCOT   SU 286995    20km NNE of Swindon

Bridge repairs here are mentioned in 1208.  The sharply pointed side-arches with ribs without chamfers look 14th century. The four-centered central arch surmounted by the base of a former cross on the parapet must date from after 1387, when the breaking of the bridge by the Earl of Derby trapped Richard II's favourite Robert de Vere, Earl of Oxford, and allowed his capture and execution by the Lords Appellant.

*Old view of the  cathedral, castle and medieval bridge at Rochester*

## ROCHESTER   TQ 740688   On the NW side of Rochester

The only relic of the new eleven-arch bridge of 1383-92 built by Sir Robert Knolles and Sir John de Cobham is the chapel of All Souls or the Holy Trinity associated with its SE end lying immediately below the castle. Originally containing a chantry in memory of de Cobham the chapel was restored from ruin in 1937. It had previously served as a greengrocer's shop until 1879. Designed by Henry Yevele, the bridge had a central drawbridge to let ships pass up the Medway and was 170m long (see p43). The previous bridge swept away in 1381 was of timber and was dangerous to cross in stormy weather since it had no handrails. It had nine piers and unequal spans. An attempt was made to burn it during the siege of 1215 and in 1343 the west end was extended with a barbican and outer drawbridge. One pier of the wooden Roman bridge was found in 1851 during construction of a new bridge further downstream, replaced in 1914.

## SALISBURY   SU 144290 & 141298 & 158297 Various locations around Salisbury

In the 1240s Bishop Bingham built Harnham Bridge over the River Avon to bring traffic from the road to Wilton into his town of New Sarum. The bridge is in two parts over separate branches of the river, the southern part being of six ribbed arches, now largely obscured by widening of 1771 on both sides, whilst Leland says the northern part had four arches. Both parts retain 13th century arches. Between the two bridges is a much altered 13th century chapel of St John set surprisingly close to water level, later used as a house. Maintenance of the bridge and chapel was vested in the hospital of St Nicholas. It was widened in the 18th century but is now bypassed by a bridge of 1933.

One 15th century arch remains in the Crane bridge, which was much widened southwards in 1898. Leland describes the Fiskerton bridge north of it as having six stone arches. It was rebuilt in 1762 and replaced by an iron bridge of 1872. The Milford bridge over the River Bourne east of the city comprises two pairs of arches lying on either side of a length of causeway. It may date from soon after 1386, when the bridge here was reported to be broken. There is a moulded stringcourse below the parapet.

## SOMERSET   SU 922439   5km west of Godalming

Three 13th century arches over the River Wey built by Waverley Abbey are divided by cutwaters and there is an extra later arch of brick at the south end.

## STEDHAM   SU 854229   3km NE of Midhurst, 11km SSW of Haslemere

This bridge has six low segmental arches across the River Rother. There are original parapets and small cutwaters on the upstream western side and small buttresses on the downstream eastern side.

## STOPHAM   TQ 030183   9km SE of Horsham

The high centre arch dates from 1822 and one of the other arches was replaced after the Civil War. The other five low plain round arches represent what was built by the Rector of Pulborough in 1423, to replace a wooden bridge over the River Rother. There are pedestrian refuges on both sides, the refuges on the south being semi-hexagonal. The bridge is now bypassed by a bridge of 1986 on the upstream northern side which only slightly spoils the surroundings of this, the best medieval bridge in Sussex. See p17.

## TESTON   TQ 709533   5km SW of Maidstone

The three slightly pointed centre arches with cutwaters carried up to support semi-hexagonal refuges are 15th century, although the parapet is modern. The side arches of this bridge over the River Medway are 19th century.

*Bridge over the Medway at Teston in Kent*

*Bridge over the Rother at Stopham in Sussex*

*This is the longer and more south-easterly of the two medieval bridges at Tilford in Surrey*

## TETBURY    ST 893930 & 889929    16km SW of Cirencester

The 16th century Wiltshire bridge is actually more of a causeway with a single small arch over the River Avon. There is a large buttress on the south side. Smaller buttresses on the north are later. The Waters bridge of 1622 lies to the west.

## THORNBOROUGH    SP 729332    3km east of Buckingham, WSW of Milton Keynes

The 14th century bridge of five arches over the Padbury Brook is now bypassed by a new bridge of 1979. The ribbed middle arch and the eastern arch are four-centred. The pointed arch between them also has ribs. Three arches have hoodmoulds. There are three refuges on the upstream southern side, whilst the northern side has a rectangular recess and a buttress, both towards the western end, and patching with blue bricks.

## TILFORD    SU 872435 & 874433    7km south of Aldershot

The monks of Waverley Abbey built the bridges over the two arms of the River Wey, each bridge having pointed cutwaters facing upstream and rounded ones facing down-stream. One lies at the north end towards Farnham and is of four arches with tiny cut-waters. A fifth arch is now hidden from view. A modern bridge lies downstream of it. The other bridge lies to the east at the bottom of the green and has seven arches, including one of brick dating from the 16th century at the southern end. See page 45.

## TROTTON    SU 836223    5km WNW of Midhurst

Probably built by Lord Camoys in the early 15th century , this is a bridge of five round arches each with five chamfered ribs over the River Rother. The cutwaters have lost their former pedestrian refuges. There is no access to the sides. Despite being unwid-ened this bridge still carries A-road traffic controlled by traffic-lights.

## TWYFORD    TQ 698500    9km  SW of Maidstone

The bridge over the River Medway has  four irregular pointed arches and brick refuges on the cutwaters. The bridge lies near the confluence of the Medway and the Beult, and the name Twyford is a corruption of "twin-ford".

## UNSTEAD    SU 993454    3km south of Guildford

The five-arch bridge over the River Wey built by the Cistercian monks of Waverley Ab-bey was largely rebuilt in the 19th century.

## WALLINGFORD    SU 610895    On the east side of Wallingford

Only five of the nineteen arches actually span the River Thames and the three central arches surmounted by a balustrade date from 1809, following flood damage to arches of 1751 built to replace arches destroyed in the 1640s and replaced by drawbridges. Arches of 1751 survive at either end of those of 1809. At the town end west of the river are a medieval arch with ribs and another arch which has been blocked up on the north side. The other arches form a causeway to the east. A ribbed medieval arch and one of 1751 lie between pairs of 16th century arches, then there is another ribbed medieval arch. Three arches at the east end are additions of 1809. The bridge was widened in 1770. A toll-house of that era survived until 1950.

## WHEATLEY    SP 612052    10km ESE of Oxford

Just one of the eight stone arches mentioned by John Leland now survives, embed-ded in the east end of a 19th century structrure. A bridge here over the River Thame is mentioned in 1286. A skirmish took place here in the Civil War.

*Twyford bridge near Yalding in Kent*

## WISBOROUGH    TQ 068260    11km WSW of Horsham

One pointed arch of c1500 and two round arches of c1700 remain under the red brick New Bridge of 1839 over the River Arun.

## WOOLBEDING    SU 873220    1km NW of Midhurst

This bridge over the River Rother has four arches each with three chamfered ribs and cutwaters. The parapet appears to have been rebuilt using original coping stones.

## YALDING    TQ 698500    8km SW of Maidstone

The 30m long Town bridge of ragstone across the River Beult has two sets of three pointed arches of the 15th century on either side of a later round central arch.

BRIDGES ASSOCIATED WITH CASTLES, TOWN DEFENCES, MONASTERIES:
Eltham TQ 425740, Herstmonceux TQ 647104, Hyde Abbey SU 475294,
Leeds TQ 836532, Michelham PrioryTQ 559093, Sandwich TR 331578

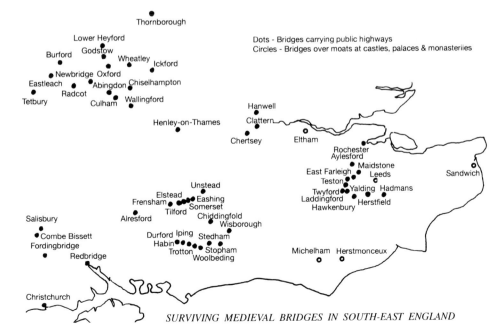

*SURVIVING MEDIEVAL BRIDGES IN SOUTH-EAST ENGLAND*

# MEDIEVAL BRIDGES OF EASTERN ENGLAND

## ALCONBURY   TL 186757   5km NW of Huntingdon

The 15th century Church Bridge has four pointed chamfered arches over the Alconbury Brook. There are buttresses on the cutwaters, one cutwater being sloped.

## BARFORD   TL 134516   9km ENE of Bedford

The bridge over the River Ouse has seventeen arches of different types. Eight of them are 15th century and have chamfered arch-rings. The rather fancy brick outer arches between the medieval cutwaters with refuges on the east side are a widening of 1874, although the bridge is still only able to carry single-file traffic. See page 21.

## BROMHAM   TL 001507   4km WNW of Bedford

The causeway and bridge over the River Ouse together have twenty-six arches, some of which are 15th century, although these are difficult for the public to see since there is no access to the sides or underneath. Some arches date from the late 15th century and others from a rebuilding of 1813. The bridge itself has six arches, including one arch over the mill-race, and has been widened on the southern downstream side. Here widening has been achieved by replacing straight sections of parapets between refuges on the cutwaters by curved sections of parapet set further out, thus reducing the size of the refuges. A chantry chapel once lay beside the bridgehead.

## COGGESHALL   TL 849223   At Coggeshall, 15km west of Colchester

There is 13th century brickwork on the west side of the Long Bridge over the new course of the River Blackwater as diverted by the monks of the abbey. There are three arches with the tallest one in the middle. The bridge was rebuilt in 1705, and widened slightly on the downstream eastern side in 1912, that date appearing on it. The original course of the river (now known as the Black Ditch) is crossed by the early 19th century Short Bridge not far to the north.

## CRINGLEFORD   TG 199059   4km  SW of Norwich

This small early 16th century bridge over the River Yare has two four-centred arches. The abutments and pier base may be older. Widening c1780 was achieved by adding extra arch-rings on either side.

## HADLEIGH   TM 025421   At Hadleigh, 14km west of Ipswich

The Toppesfield bridge is a 14th century brick structure of three pointed arches over the River Brett. Just one arch ring with a chamfer is of stone. The brick ribs with chamfers are a great rarity. The north-west upstream side has been widened.

## HARROLD   SP 955566   South of Harrold, 12km NW of Bedford

The bridge has six arches over the River Ouse and nine more in the causeway to the south. The bridge arches are widely and irregularly spaced and semi-circular except for a taller and finer pointed one with two arch-rings of ironstone. There are cutwaters without refuges on the upstream western side The upstream side of the causeway is curved out into a series of cutwaters following modern widening with brick arches set in front of the small medieval ones. There are no parapets on the causeway but a hand-rail now protects the side facing the roadway below it. See pages 13 and 51.

## HEIGHAM   TG 420185   Near Potter Heigham, 15km NW of Great Yarmouth

The round central arch of the bridge over the River Thurne may be later and has probably taken the place of two original arches, but the pair of pointed side-arches with chamfered ribs are probably of the 1380s.

*Coggeshall, Essex, an early brick bridge*

*Bishop's Bridge at Norwich*

*Hadleigh, Suffolk, with unusual brick ribs*

*Bury St Edmunds Abbey: Abbot's Bridge*

*Alconbury Bridge, near Huntingdon*

## HINCHINGBROOKE   TL 225712   1km SW of Huntingdon

Named after nuns of the nearby priory, the Nuns Bridge over the River Ouse has pointed arches on either side of a wider elliptical arch, and two 18th century western arches.

## HUNTINGDON   TL 243715   SE of Huntingdon

The fine six-arch bridge over the River Ouse was referred to as "lately built" in 1332. One arch of the bridge destroyed in 1294 by flooding caused by a build-up of ice, may remain at the north end, widened with the aid of a rib on each side. There is arcading on the upstream SW side over the next two arches. A cutwater between them on the NE side carried a chapel of St Thomas Becket. The other three arches seem to have been the work of a different authority, with polygonal refuges. That nearest the middle of the bridge has been rebuilt as a drawbridge replaced it in the 1640s. See page 20.

## KING'S LYNN   TF 616200 & 623203   To west and NE of King's Lynn

The single wide arch over the Purfleet near the Customs House was mostly rebuilt in the 19th century but some brickwork of the 17th century and earlier survives. The medieval bridge had a chapel on or beside it.  Near the surviving section of the town wall there survives a medieval arch with chamfered arch-rings over the Gaywood River.

## MAYTON   TG 250216   13km north of Norwich

The brick bridge over the River Bure has two four-centred arches with slight chamfers. Each parapet has at the east end a niche with a seat under a roof.

## MOULTON   TL 698646 & 698644   At Moulton, 5km east of Newmarket

A hump-backed 15th century bridge of four arches (now an ancient monument in the care of English Heritage) lies beside a ford on the River Kennet. The brick arches have been somewhat restored, although original parts with a chamfer remain on the south side.  Further south lies a second 15th century single-arch bridge built of flint.

## NEWTON FLOTMAN   TM 213980   11km south of Norwich

Three arches across the River Tas have chamfered ribs. A fourth arch was rebuilt in brick in 1835 and a fifth arch was added as part of later 19th century widening.

## NORWICH   TG 240090   On the east side of Norwich

Bishop's bridge over the River Wensum has arches of brick and stone and a shield on the downstream side of the central arch. The western cutwaters support corbelled, truncated bases of two circular turrets which flanked a gateway built in 1343 at the expense of Richard Spink and demolished in 1791. The Whitefriars' bridge was destroyed by the flood of 1290 and demolished in 1549 to provide timber for shoring up gates against the rebel attack. It was not rebuilt in stone until 1591, and is now a 20th century structure. The Fye Bridge first mentioned in 1153 was rebuilt in stone in the early 15th century. It was replaced in 1572 by a two-arch structure, repaired in 1756 and replaced in 1933-4. The Blackfriars' bridge was of wood until a three-arch stone bridge was built in 1572.  The St Miles Bridge, Coslany is of 1804 replacing a stone structure of 1521.

## ST IVES   TL 312712   On south side of St Ives, 7km east of Huntingdon

The bridge over the River Ouse dates from c1415 and retains original cutwaters on both sides. Projecting from the middle of the east side is a chapel of St Lawrence with a polygonal  apse and a room underneath it. The roof and parapet date from the 1930s when two upper storeys of brick added later for domestic use were removed.  The two southern arches of the bridge were rebuilt in 1716. See picture on page 15.

## SPALDWICK   TL 127730   At Spaldwick, 10km west of Huntingdon

The middle arch of three across the Ellington Brook has three ribs. Both arch-rings are chamfered. The bridge has been widened on the upstream side. It has brick parapets.

## SUTTON   TL 221474   NE of Biggleswade, 17km east of Bedford

There is a ford downstream of this 14th century bridge of two arches with a refuge between them on the upstream northern side. The top has been lowered since the parapets cut into the uppermost arch-ring of each arch. The parapets look like reused late medieval material perhaps taken from a suppressed monastery and the bridge may originally have been without any. Timbers of an earlier bridge have been found here.

## TURVEY   SP 938524   SW of Turvey, 11km WNW of Bedford

The 200m long causeway and bridge over the River Ouse together have sixteen arches and cutwaters on the upstream side only. Some of the eight pointed arches may be 13th century. One arch is segmental. Other arches are of 1795 and the 1820s.

## WIVETON   TG 044426   East of Blakeney and 13km west of Sheringham

The ribbed arch over the River Glaven has a span of 9.5m. A chapel stood to the south of the western abutment and originally there was another arch to the west.

OTHER BRIDGE: RUSHFORD TL 925812   Possibly 16th century but mostly rebuilt.
BRIDGES AT MONASTERIES: Bury St Edmunds (see p49), Walsingham & Waltham
MOAT BRIDGES: Attleborough Hall TM 051963, Bletsoe Castle TL 025584,
    Earl Soham Lodge TM 232635, Framlingham Castle TM 287637, Hedingham
    TL 787358, Hunstanton Hall TF 692418, Latchleys TL 671396, Park Farm
    TL 360776, New Buckenham Castle TM 084904, Pleshey Castle TL 665145

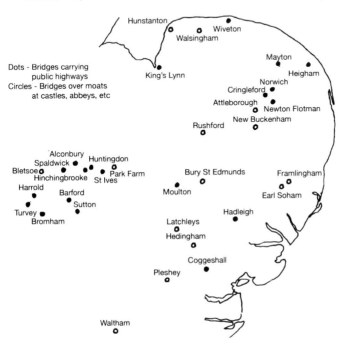

*SURVIVING PRE-REFORMATION BRIDGES IN EASTERN ENGLAND*

*Causeway with cutwaters at Harrold, Bedfordshire*

# MEDIEVAL BRIDGES OF THE MIDLANDS

ANSTEY    SK 553084  &  556089    5km NW of Leicester

Parallel to the nearby B5322 Leicester Road is a packhorse bridge 16m long and 1.5m wide between the parapets. There are five roughly made arches over the Rothley Brook. Also crossing that watercourse is the widened King William's bridge with two round arches and a cutwater at the end of Sheepwash Lane. Jervoise suggested a late 17th century date for both bridges but the southern one may be late medieval. It has a cobbled roadway and cutwaters with refuges.

*Bridge over the River Derwent at Bakewell, Derbyshire*

*Banbury, Oxfordshire*

*Packhorse bridge at Anstey, near Leicester*

*The bridge over the River Avon at Bidford. Notice the brick patching on the cutwaters*

**ASFORDBY**   SK 706186   5km west of Melton Mowbray

A packhorse bridge of three pointed arches crosses the River Wreak south of the village. It has been widened in brick.

**AYLESTONE**   SK 568009   4km SW of Leicester

At the west end of Marsden Lane a series of eleven small 15th century arches of various sizes totalling 50m long form a bridge and causeway over the former line of the  River Soar, now canalised and further east. Two of the upstream southern cutwaters and one on the north have refuges of various shapes and sizes. See plan on page 17.

**BAKEWELL**   SK 219687   NE side of Bakewell, 11km NW of Matlock

There is a base of a cross on the parapet on the south side of the bridge of five arches over the River Wye. Each arch has four original ribs and four more provided when the bridge was widened on the north side in the 19th century.

**BANBURY**   SP 462406   On the east side of Banbury.

A water mill and an adjacent bridge over the River Cherwell are mentioned in 1294. The bridge seems to have been rebuilt c1500-40 and had then had four arches, but by the 18th century it had seven pointed arches. Just two medieval arches now remain, embedded between modern widening in brick on both sides.

**BASLOW**   SK 252723   6km NE of Bakewell

The bridge across the River Derwent is dated 1608 and has a toll-booth in the NW corner from the time of the 1649 repairs but the three arches each with six ribs probably go back to the late 15th or early 16th century.

**BELGRAVE**   SK 591074   3km north of Leicester

This bridge over the River Soar is mentioned in 1357, when barriers under it were ordered to be removed, and again c1540 by Leland. A roadway originally just 2.4m wide was carried on seven arches. There has been widening on both sides with brickwork.

**BEWDLEY**   SO 787753   On the east side of Bewdley,  WSW of Kidderminster

Richard III donated 20 marks towards building a new bridge here over the River Severn in 1483. A view made before the bridge was destroyed by flooding in 1775, and then replaced by a Telford bridge of 1781 further upstream, shows wooden decking across two broken arches and a cottage towering over one pier which had formerly borne a gatehouse. The two end abutments still remain, that on the town side only being rediscovered by an excavation under the site of a former bandstand in 2004.

**BIDFORD**   SP 099517   17km SSE of Redditch

This eight-arch bridge over the River Avon is a composite structure. The two northern arches and the most southernly one have double arch-rings of thin slabs and look 13th or 14th century, although a joint in the southern arch suggests widening, perhaps during construction. John Leland mentions stone from recently dissolved Alcester Priory being used to repair the bridge in the mid 16th century, and this could refer to the second and third arches from the south end. The low fourth arch and the high semicircular fifth arch could represent repairs made in 1650 after damage caused during the conflicts of 1644. The upstream eastern side has cutwaters with refuges much patched in brick. The downstream side has been slightly widened, giving a width between the parapets of 3.9m. There is an additional flood arch some way further south.

*An unusually finely moulded arch at Cromford, Derbyshire*

*Bridge chapel at Derby*

## BRAYBROOKE   SP 765845     At Braybrooke, 4km SE of Market Harborough

This bridge of brown stone over the tiny River Jordan was begun c1400 by Sir Thomas Latimer. There are three pointed and chamfered arches with an added cutwater on the upstream side facing east, and low parapets.

## BRETFORD   SP 429769     10km east of Coventry

Four pointed arches over the River Avon plus a segmental arch at the north end are medieval but are now encased within 18th century widenings on each side.

## CASTLE DONNINGTON   SK 457302   15km SE of Nottingham

The remains of a 13th century stone bridge which once carried the Derby to London road over the former course of the River Trent were found in a gravel pit in 1993. About 50m NE of them timbers of a late 11th century wooden bridge were also found.

## CHARWELTON   SP 535561   7km SW of Daventry

This is a packhorse bridge of two small pointed arches over the River Cherwell. There is a single cutwater on the upstream northern side, facing the A361 bypassing it. There are finely moulded parapets suggesting a 15th century date.

## COLESHILL   SK 199896     North of Coleshill, 13km east of Birmingham

The River Cole to the north of the village is crossed by a 16th century bridge of six sandstone segmental arches with cutwaters and double arch-rings, both chamfered. The upstream western side has been widened in brick in the 18th century. Three flood arches in a causeway to the north may also be of that period.

## COLLYWESTON   SK 990036   5km SW of Stamford

This bridge over the River Welland has three pointed medieval arches with the innermost of two arch-rings recessed, and three segmental arches dating from 1620.

## COTES   SK 553206   1km NE of Loughborough

The piers and one arch at the west end of the bridge over the River Soar are medieval. The other arches are 19th century work of brick.

## CROMFORD   SK 300571     At Cromford, 3km south of Matlock

Adjoining the south end of this 15th century bridge, of three four-centred arches with unusually fine moulded arch-rings over the River Derwent, is a ruined chapel of the same era. The upstream western side of the bridge has been widened.

*Trinity Bridge at Crowland in Lincolnshire*

*Darley Bridge, Derbyshire*

### CROWLAND   TF 240106   In Crowland, 12km south of Spalding

The triangular bridge now only has dry land beneath the three arches meeting at 120 degrees but originally there was a meeting here of waters of the Nene and Welland. The arch ribs are moulded with hollow chamfers and wavy chamfers, suggesting a late 14th century date, but a triangular bridge here is mentioned as early as 943. It served only as a footbridge, being reached by three steep flights of steps and seems to have additionally formed the base of a tall cross. The large seated figure of Christ is 13th century and probably from the west front of the nearby abbey. See plan on page 17.

### DARLEY   SK 270620   3km NW of Matlock

This is a 15th century bridge of five arches over the River Derwent. In 1682 the bridge had seven arches, one of them filled up with silt. Two sharply pointed arches at the east end each have four plain ribs. The other arches are segmental. The bridge has been doubled in width on the northern upstream side.

### DERBY   SK 354368   To the SE of Derby city centre

The springing of an arch of the old bridge can be seen under the east end of the small chapel of St Mary first mentioned in 1328. There is a squint towards the altar from the outside of the north wall, where the roadway lay. The east window is late medieval and there is a half-timbered gable above. The building was later used as a pair of cottages and then as a carpenter's shop before being restored to use as a chapel in 1873. The south windows date from a restoration in 1930.

### DITCHFORD   SK 930684   4km east of Wellingborough

Six semi-circular arches each with two chamfered arch-rings cross the River Nene. The parapet is corbelled and has five pedestrian refuges on cutwaters on each side. One refuge on the west side near the north end bears the cross keys of Peterborough Abbey, which built the bridge in the 14th century. At each end is a causeway with three further arches. An eight-spoked wheel of St Catherine appears on the other side.

### DUDDINGTON   SK 985009   West of Duddington, 7km SW of Stamford

Beside the 14th century bridge over the River Welland is a restored watermill dated 1664, recently used as offices. The four arches of the bridge are pointed and have chamfered arch-rings facing upstream. They were widened with blue bricks and an ashlar face with keystones in 1919. A bridge here was described as broken in 1380.

**DUFFIELD**    SK 350430    SE of Duffield,  8km north of Derby

Church bridge over the River Derwent has two pointed 17th century arches and a 16th century four-centered arch to the west with finely moulded edges. The bridge was widened on the upstream northern side in 1802.

**ECKINGTON**    SO 922422    15km SE of Worcester

This is a narrow sandstone structure with cutwaters with small  pedestrian refuges. Mentioned in 1440 and 1573, it was ruinous in 1634 and was much rebuilt in 1728.

**EGGINGTON**    SK 268270    4km NE of Burton-upon-Trent

Between the brick aqueduct carrying the Trent and Mersey Canal over the River Dove and the wide new bridge of the A38 is the 15th century Monks' Bridge carrying the Roman road called Ryknild Street over the river, and named after the Benedictine monks of Burton Abbey. The SW arch was rebuilt in 1695 and the bridge was widened by 2m on the NW or upstream side in the 19th century. As a result the cutwaters now only carry refuges on the downstream side.  The three original arches each have three ribs, the outer ones having chamfers on their outermost sides only.

**ELFORD**    SK 191099    8km north of Tamworth

Crossing a former channel of the River Tame, this bridge of seven round arches is probably 16th century. Refuges flank the higher central arch.

**ENDERBY**    SP 550985    SE of Enderby, 7km SW of Leicester

This is a 15th century packhorse bridge of two pointed arches with chamfered arch-rings and cutwaters. Originally a crossing of the River Soar, it now lies in a field.

**ESSEX**    SJ 995225    West of Great Haywood, 7km east of Stafford

Fourteen arches carry a pathway just 1.2m wide over the combined waters of the River Sow and River Trent to the east of Shugborough Hall. All the piers have pedestrian refuges on the cutwaters on each side. The SW end of the bridge has a bend in it. It is probably mid 16th century, and is the longest packhorse bridge in England. See p17.

**EVERDON**    SP 599574    6km SE of Daventry

A tributary of the River Nene is crossed by two pointed arches each with three ribs. There is a refuge on the cutwater facing upstream to the west, but only a pilaster buttress on the east side.

**FURNACE END**    SP 248912    17km ENE of Birmingham

The road towards Shustoke crosses the River Bourne with one medieval arch with four chamfered ribs. A datestone commemorates a second widening to the south in 1924.

**GEDDINGTON**    SP 894829    4km NE of Kettering

Three pointed 13th century arches and one round rebuilt arch of 1784 constitute this bridge beside a ford of the River Ise lying just south of the best surviving Eleanor Cross. Over the second arch from the north on the east side is a section of string course. Part of the head of a flood arch remains visible further south on that side. See page 17.

**GRENDON**    SK 285010    8km SE of Tamworth

Four chamfered and pointed arches with hoodmoulds and cutwaters take a track over the River Anker. The bridge was probably built by nearby Polesworth Abbey in the 15th century and has refuges on the central cutwaters.

*Packhorse Bridge formerly across the River Soar at Enderby, near Leicester*

*Essex Bridge over the River Trent, Staffordshire*

*Eckington Bridge, Worcestershire*

*The main two arches of the bridge at Marton, Warwickshire*

*Irthlingborough Bridge over the River Nene, Northamptonshire*

## HALFORD   SP 259454   11km SE of Stratford-upon-Avon

This structure is more in the nature of a causeway with three widely spaced pointed arches across the River Stour. The southern arch is chamfered. Beyond it is a fourth modern arch of brick. A modern bridge carrying A429 lies further east.

## HAMPTON-IN-ARDEN   SP 213801   16km SE of Birmingham

A track leading SE from Hampton-in-Arden towards Bradnock's Marsh crosses the River Blythe on a packhorse bridge with three 15th century pointed arches and two later segmental ones of brick. One pier towards the SE bears the base of a cross upon it.

## HUNNINGHAM   SP 373685   North of Hunningham,  7km NE of Leamington Spa

The bridge over the River Leam has a segmental west arch probably of 1651 when £20 was spent on repairs, but the two round arches are older. The east arch has two flush arch-rings facing upstream but the lower arch-ring is recessed on the other side.

## IRTHLINGBOROUGH   SP 956706   East of Irthlingborough, 12km SE of Kettering

This bridge with ten arches over the River Nene is 14th century and is now superseded by a new bridge alongside it despite being widened slightly out onto the cutwaters on the SW side c1830. Only the cutwaters on the NE side ever had refuges. Most of the arches are pointed and the middle arches have widely-spaced chamfered ribs. The cross keys of Peterborough Abbey appear on a panel on a cutwater on the NE side, where the fourth cutwater from the SE end bears the date 1668. It was probably then that the third arch from the SE end was rebuilt higher than before. Causeways continue the bridge at each end, with two flood-arches at the SE end, and three towards the NW. John Pyel of London left funds towards work on this bridge in his will dated 1382.

*Bridge over the River Derwent at Matlock in Derbyshire*

*The High Bridge at Lincoln*

*Hampton-in-Arden, Warwickshire*
*pedestal for a cross or statue*

*Bridge at Medbourne*

## KETTON    SK 982043    Near Ketton Church, 5km SW of Stamford

This bridge over the River Chater in Rutland has three pointed arches. It was widened on both sides with medieval-looking hoodmoulds in 1849, that date appearing on it.

## LINCOLN    SK 975711    In the middle of Lincoln

Just south of the late medieval Stonebow gateway lies the High Bridge over the River Witham, a semicircular arch spanning 6.7m, with five ribs including unusual diagonal ones. It is 12th century origin but much restored. Four extra arches are no longer visible. An added projection of 1233 on the east side carried a chapel of St Thomas Becket which was replaced in 1762 by an obelisk, now itself gone. The other side bears a row of timber-framed shops on a 16th century widening. They were mostly rebuilt in 1900, when both sides of the bridge were refaced. All the windows project like oriels.

## MARTON    SP 407691    11km SSE of Coventry

According to Dugdale this bridge over the River Leam of two depressed pointed double-chamfered arches dying into the abutments and cutwaters was built by John Middleton in 1414 to succeed a wooden toll-bridge. There is a third arch further west. An older bridge here was mentioned in 1234. The bridge was widened in 1926 but is now bypassed by a new bridge of 2000 and has been reduced to its original width.

## MATLOCK    SK 298601    In the centre of Matlock

The fine 14th century bridge over the River Derwent with four pointed arches with sunk-quadrant mouldings and large cutwaters was much widened on the northern upstream side in 1904, where there are refuges. There are hoodmoulds over the arches.

## MAYFIELD    SK 158458    3km SW of Ashbourne

Widening in concrete faced with sandstone in 1937 has largely obscured the five pointed 14th century arches with sunk chamfers. Captured Scotsmen from Bonnie Prince Charlie's retreating Jacobite army are said to have been hanged from the bridge.

## MEDBOURNE   SK 799931    8km NE of Market Harborough

West of the church is a bridge of four chamfered round arches with wooden handrails over the Medbourne Brook. Cutwaters on the east side are later additions and squinches suggest parapets existed or were intended. Possibly 16th century, the bridge allowed access to the churchyard, although it is usually described as a packhorse bridge.

## MILLDALE   SK 139546    Near Alstonefield, 10km NNW of Ashbourne

The small packhorse bridge known as Viator's Bridge has a cutwater between two early 16th century four-centred arches with hoodmoulds over the River Dove. It is mentioned in Izaak Walton's book The Compleat Angler, first published in 1653.

## NEW MILLS   SK 002859   At New Mills, 13km SE of Stockport

A late-medieval packhorse bridge of two arches over the River Sett has been widened and altered with concrete skews to allow a road to cross it at an angle.

## NOTTINGHAM    SK 582381 &   568395   1.2km SSE of Nottingham Castle

A bridge is mentioned here in 924. The new bridge begun in 1156 had over twenty arches, and a chapel of St James founded in 1303 by John le Palmer at the south end of its causeway, the total length being about 150m. The bridge was described as broken in 1335. A photo of 1871 shows the downstream side of the old bridge (which had been later widened on the upstream side) and the new bridge begun in 1868 nearby to it, after the old bridge was discovered to have inadequate foundations. Two pointed 14th century arches with triple arch-rings of the causeway to the old bridge survive in a traffic roundabout on the SE bank of the River Trent.  A medieval bridge also survives over the dry moat in front of the outer gatehouse of Nottingham Castle.

## OAKAMOOR   SK 054449   East of Stoke on Trent

One four-centered arch remains at the SW end. Three segmental arches over the River Churnet are 18th century, when the bridge was widened on the upstream NW side.

## OUNDLE   TL 037878    SW of Oundle,  15km SW of Peterborough

The three segmental northern arches over the River Nene appear to be medieval. The southern arches are 18th century. There are two cutwaters on the upstream west side.

## PERSHORE   SO 952451    South of Pershore, 14km SE of Worcester

The bridge over the River Avon has two parts, both widened, one of three segmental arches with cutwaters and the other of one large later central arch and five smaller arches with cutwaters only on the upstream eastern side. The bridge is first mentioned in the late 13th century. Leland saw stone being brought here for repairs c1540 but the bridge was ruinous again by 1607 and it was mostly rebuilt after Civil War damage.

## POWICK   SO 835523    3km SW of Worcester

The 15th century bridge with three skewed arches of sandstone now only carries a cycleway over the River Teme, since an iron-arched bridge of 1837 now carries the A449 just to the east. An older bridge here was reported to be a dangerous condition in 1336 and the Prior of Great Malvern was obliged to repair it. The existing bridge was decayed in 1598 and in 1633 Sir Thomas Bromley agreed to have it repaired. Here in 1642 was fought a battle between Royalists and Parliamentarians. The Royalists made the bridge unusable and in 1645 Oliver Cromwell used a bridge of boats to get his army across the River Teme. Some of the arches appear to be a rebuild of after that period.

## REARSBY    SK 652146    9km NE of Leicester

Six semicircular arches carry a roadway just 1.5m wide beside a ford over a brook south of the church. There are round cutwaters on the upstream side. See photo, page 4.

## ROWSLEY    SK 257658    7km NW of Matlock

The bridge carrying the A6 over the River Derwent was considerably widened in 1925 on the upstream NW side. The older part is 15th century and has five ribbed and pointed arches and pyramidal-topped cutwaters.

## SCREDINGTON    TF 097409    9km SSE of Sleaford

The pair of four-centred arches of good ashlar over the North Beck are of late medieval origin but were rebuilt in the 20th century, when the pier between them was given an ugly concrete base. The upstream side has been widened and the other side refaced.

## SHIPSTON-ON-STOUR    SP 261405    15km SE of Stratford-upon-Avon

This 16th century bridge has five four-centred arches over the River Stour. One segmental arch is probably 17th century. The bridge bears the date 1698 and has been widened on the upstream southern side.

## STAMFORD    TF 030069    On the south side of Stamford

This bridge over the River Welland is mentioned in Domesday Book and was a five arch structure of stone by the late 12th century. A hospital chapel of St John the Baptist lay beside the SE bridgehead. Until it was demolished c1778 the upper room of the gateway at the NW end was used as the Town Hall. Most of the bridge was replaced by a new structure in 1848-9 but one arch remains mostly hidden away on the SE side.

## STARE    SK 329715    4km east of Kenilworth,  7km south of Coventry

The bridge has three pointed arches over the River Avon and a causeway with six further arches, two of which are segmental. On the downstream side there are pairs of thin buttresses between each arch. The upstream eastern side has cutwaters with refuges, the first cutwater from the south end, standing midstream, being one of the largest in England with a width of 6.5m.  A new bridge now carries A444 further east.

## STONELEIGH    SP 332727    6km south of Coventry

The bridge has three arches over the River Stowe and five flood arches to the  west. The medieval north side has a triangular cutwater. The south side of the early 19th century  has half-round cutwaters.

*Remains of Trent Bridge, Nottingham*

*Stare Bridge, Warwickshire*

*Clopton Bridge at Stratford-upon-Avon*

## STRATFORD-UPON-AVON   SP 205548   Close to the SE of town centre

The River Avon at Stratford is crossed by a bridge of fourteen flattened pointed arches named after Sir Hugh Clopton who built it in the 1480s to replace a wooden structure known to have had an associated chapel and hermitage by the 1440s. Sir Hugh was Lord Mayor of London in 1492 and was buried in a chantry chapel in Holy Trinity church in 1496. Two arches were rebuilt in 1524 and there was further rebuilding after the 1588 floods broke arches at each end. Another arch had to be replaced after being destroyed during the conflicts of the 1640s, and the very low parapets were raised in 1696. There were once five more arches at the north end. The tower-like toll-house was added in 1814, following widening and the replacement of the third pier from the east end.

## SWARKESTONE   SK 370278   8km south of Derby

Five segmental arches of 1795-7 carry A514 over the River Trent. South of it the road is carried on a causeway with seventeen pointed and ribbed flood-arches, most of them probably 14th century, but widened to the east in 1853, and some underpinned with brick in 1899. Others were replaced in 1802. Originally the bridge and causeway together had thirty-nine arches. The bridge is mentioned in 1275 and 1327 and had a chapel at the south end of the river crossing in 1558.

## TEWKESBURY   SO 894333   At the north end of the town centre

This bridge crosses the River Avon near its confluence with the River Severn. Two round arches each with four large plain ribs with slight chamfers may go back to the time of King John (ie c1200) after whom the bridge is now named, although it was called the Long Bridge in the 19th century. South of them is a higher navigation arch of 1824 and then a third medieval round arch with ribs. Another modern arch lies to the north. The bridge was considerably widened in 1962. A causeway links it with another bridge of three arches, now all of the 1962 rebuild. This northern part was recorded as broken in 1367 and one of its arches was rebuilt in 1748. See picture on page 64.

## THORPE WATERVILLE   TL 023813   NE of Thrapston, 15km SE of Corby

This bridge over the Thorpe Brook has a 14th century ribbed arch but has been widened on both sides in the late 18th century.

## THRAPSTON   SP 990786   West of Thrapston, 14km SE of Corby

Four of the nine arches over the River Nene have double-chamfered arch-rings. The other five were rebuilt in brick after the flood of 1795 and then widened later. A bridge here is mentioned in 1224. The bridge had eight arches when seen by Leland c1540.

## TIDMINGTON   SP 261382   & 245375  17km SSE of Stratford-upon-Avon

One pointed arch is said to remain of the bridge over the River Stour existing by 1615. The round arches are 18th century, when the bridge was much widened. Of greater interest is the packhorse bridge to the west with one pointed arch and one round one.

*The much-rebuilt bridge over the Teme at Powick, near Worcester*

*At Swarkestone in Derbyshire the bridge itself has been replaced but an impressive causeway still remains*

*The bridge at Water Orton in Warwickshire*

*King John's bridge at Tewkesbury*

*Wakerley: reset head on upstream side*

## UTTERBY   TF 305992   6km WNW of Louth

This tiny 14th century bridge in the middle of the village has a double-chamfered arch. There are no parapets.

## UTTOXETER   SK 106344   1km NE of Uttoxeter, 17m NW of Burton-upon-Trent

Two arches of the Dove bridge appear to be mid 14th century work. The arches have chamfered arch-rings and one has four ribs. The date 1691 on one of the parapets refers to the building of the two semicircular central arches. The bridge has been widened on the upstream side. Just below the bridge the River Tean joins the River Dove, the latter forming the boundary between Staffordshire and Derbyshire.

## WAKERLEY   SP 951999   10km SW of Stamford

The five pointed arches of this 14th century bridge across the River Welland have double arch-rings with both rings chamfered. One arch has carved heads over each side of it, one having been reset in 1793 when the bridge was widened on the upstream side.

## WALTON-ON-TRENT   SJ 902334   Just to the SW of the town of Stone

Two medieval pointed arches with plain ribs survive of this bridge over the River Trent. The medieval work is encased within 18th century arches with pilasters between them. There are one large and three small flood arches to the north and one to the south.

## WARWICK   SP 286647   Below the castle, on the SE side of Warwick

Mill Street is now a dead end but originally linked up with a medieval bridge over the River Avon of which two ruinous and overgrown arches still stand. The Buck brothers' print shows six arches but John Leland mentions as many as twelve. An older bridge here was said to be broken down in 1374.

## WATER ORTON   SK 174915   11km ENE of Birmingham

The sandstone ashlar bridge across the River Tame built in c1520 by John Harman, Bishop of Exeter has six round arches with chamfers and a road 3m wide drained by holes between large and closely set cutwaters. An older bridge here is mentioned in 1459. The roadway has been raised and the parapets rebuilt. See page 63.

## WEST RASEN   TF 063894   19km NE of Lincoln

West of the church lies a packhorse bridge of three segmental arches over the River Rase. Each arch has three ribs and there are cutwaters on the upstream side. Bishop Dalderby may have built this bridge in the 14th century. It now only leads to a house.

## OTHER BRIDGES PROBABLY OLDER THAN c1600 (not on the map below)

ASHFORD  SK 194694   One of the three old bridges could be 16th century in origin.
ASHOVER  SK 350626  Low clapper bridge over River Amber south of Ashover.
BACONS END  SP 183874  Pointed arches may be medieval. Others round. Widened.
BINTON SP 145530  North: seven arches of 1783. South: 1804 with medieval parts.
BISHOP'S ITCHINGTON  SP 404556   Three segmental arches over the River Itchen.
BOTTESFORD  SK 806391  c1580  Two arches, each with three chamfered ribs.
BRANSFORD  SO 804532 Submerged pier of old bridge in R. Teme by 1926 bridge.
BUGBROOKE  SP 674567   Damaged clapper bridge by canal 8km SW Northampton
EDALE  SK 088861  Single arch bridge with pathway just 0.7m wide over it.
ILAM HALL  SK 132506   St Bertram's Bridge over River Manifold in grounds of hall.
LADYBOWER  SK173895   Re-erected at north end of Ladybower Reservoir.
OSBOURNBY  TF 074379   The abutments  are the only medieval parts of the bridge.
OVERSLEY  SP 094569   Six segmental arches over River Arrow. Dated 1600.
PERRY BARR  SP 071919  Four arches over River Tame. Mostly an 18th cent rebuild.
SHELL  SO 951596   Small bridge by ford on the Bow Brook, 7km SE of Droitwich.
TURTLE  SP 928985  Three widened brick arches over R. Welland. Pier bases older.
WANSFORD  TF 073992  Twelve arches. R. Nene. Mostly of 1577 & 1672-4 and later.
WISTOW  TL 280812  Three round arches. Cantilevered modern widening on top.
MEDIEVAL BRIDGES ASSOCIATED WITH ABBEYS, CASTLES, MOATED HOUSES:
Astley, SP 312895, Broughton SP 418383, Chorley Old Hall SJ 838782, Eccleshall
SJ 828296, Goodrest  SP 273689,  Harvington SO 878744,  Thornton TA 118190
Except for Thornton, these bridges are marked on the map below.

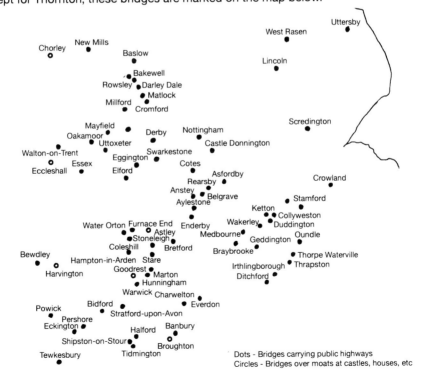

Dots - Bridges carrying public highways
Circles - Bridges over moats at castles, houses, etc

*SURVIVING PRE-REFORMATION BRIDGES IN THE MIDLANDS*

# MEDIEVAL BRIDGES OF WALES AND THE MARCHES

## ABERGAVENNY    SO 292139    1km WSW of Abergavenny

Three pointed arches at the northern end may be part of the bridge over the River Usk built by Jasper Tudor in the mid 15th century. The southern arches could be a 17th century rebuild. There are small cutwaters but no refuges. In 1811 a tramroad bridge was built close alongside on the upstream side and c1868 a wide new road-deck was built across both the bridges.

## BANGOR-IS-Y-COED    SJ 387454    8km SE of Wrexham

The five arches over the River Dee each have three arch-rings. The rectangular cutwaters may be part of the repairs carried out in 1658 after damage done in the Civil War. The road width is only 3.3m and A525 now bypasses the bridge.

## BETWS-Y-COED    SH 792567    At Betws-y-coed, 25km south of Llandudno

The Pont-y-Pair has piers set on bare rock carring three main arches plus a smaller fourth arch over the tumbling River Llugwy. The bridge supposedly dates from c1468 when the mason in charge is said to have died before it was completed, but it has been much rebuilt and was widened in 1810. The arch-rings are recessed.

## BRIDGEND    ST 904798    Near the centre of Bridgend

Following flood damage in 1775, the western two arches out of four across the Ogmore River were replaced by one span. The eastern arch became buried in buildings and was forgotten about until recently. The only pier still standing in the river bears refuges.

## BRIDGNORTH    SO 719929    Below Bridgnorth, 20km WSW of Wolverhampton

Much of the bridge now dates from 1823 but one medieval arch remains at the west end. Three of its five ribs are original. Two other arches also predate the rebuilding, An illustration of 1774 shows the second arch from the west being rebuilt. There are several mentions of pontage in the 14th century and a stone bridge is mentioned in 1478. Leland describes the bridge as being of eight arches. Until 1823 one of the piers bore a gateway with an upper room used as a chapel of St Osyth, St Clement and the Holy Trinity, first mentioned in 1404, when John Playden had it rebuilt. It was then a possession of the Franciscan friars and the offerings formed a vital part of their income.

*A widened medieval arch at Bridgnorth*     *Devil's Bridge: three superimposed bridges over a gorge*

*The bridge at Clun, Shropshire*

## CHESTER   SJ 407658   On the south side of Chester city centre.

The earliest parts of the bridge over the River Dee probably are after 1280, when it was recorded as broken. Before widening in 1826 it had a parapet only on the downstream west side and just a handrail on the other side. It now has iron handrails. Some of the cutwaters support refuges. The two segmental northern arches are modern, then there are two 14th century pointed arches with ribs. The fifth arch is similar but without ribs. The sixth arch is ribbed and semicircular and the seventh arch is pointed on the upstream side and segmental on the other side. Between these last two arches was a very substantial outer gatehouse near the south end which was being completed in 1407. It had a drawbridge in front of it. Its construction followed a period when the bridge had lain broken since 1353. The south gate of the city wall as extended in the 12th century lay at the northern end of the bridge. See pages 15 & 69.

## CLUN   SO 300808   On the south side of Clun, 22km WNW of Ludlow

A saddleback-shaped bridge of four segmental medieval arches with cutwaters plus a fifth arch of the 18th century carries the A488 linking the village on the north bank of the River Clun with the ancient parish church on the south bank.

## DEVILS BRIDGE   SN 742771   16km ESE of Aberystwyth

Three superimposed bridges span the gorge of the Mynach. The segmental middle arch with cast-iron rails in pierced panels is of 1753 and the topmost bridge is a late 20th century replacement of one of 1901. The lowest bridge could be late medieval but appears to have been refaced in the 16th or 17th century.

*Farndon Bridge over the River Dee, here forming the boundary between England and Wales*

## FARNDON or HOLT   SJ 411544    9km NE of Wrexham, 12km south of Chester

This bridge across the River Dee, here forming the boundary between Holt in Wales and Farndon in Cheshire, was built c1338-45 by John de Warenne, Earl of Surrey and Lord of Bromfield on the Welsh side. His castle of Holt lay further upstream and did not overlook the bridge. One of the piers carried a gatehouse which survived in ruins until at least the 1760s. A higher superimposed arch (see picture on page 67) marks the location of a former drawbridge in front of the gateway, which was closed from the Welsh side against an approach from Cheshire. This side of the gatehouse bore a statue of a lion whilst a crucifix faced the Welsh side. The bridge currently has eight segmental double-chamfered arches but may have once had extra arches on the Welsh side.

## HEREFORD   SO 508596   SW of Hereford Cathedral

A timber bridge to replace the ford over the River Wye from which Hereford took its name must have stood here by the 12th century. Four of the six arches of the existing bridge of sandstone now only carrying local traffic date from c1490 and are four-centred with chamfer of both arch-rings. The third arch from the north is thought to have been rebuilt after being broken as a defensive measure just before the siege of 1645, athough, rather oddly, this is the only arch to have any ribs and is lower than the others. The southernmost arch was rebuilt in the 18th century, when a gatehouse was removed from adjacent piers. The superstructure was slightly widened on both sides in 1826 with extra arches connecting the cutwaters. See front cover picture.

## KIDWELLY   SN 407069    At Kidwelly, 11km NW of Llanelli

This bridge over the Gwendraeth Fach originally had refuges on the cutwaters before 20th century widening. The two arches of differing shapes may represent different centuries, either 14th and 15th, or 15th and 16th century.

## LECKWITH   ST 159752   2.5km SW of Cardiff Castle

The middle arch of this medieval bridge over the River Ely has been rebuilt as a round arch. The surviving original arches are pointed and have recessed inner arch-rings. There are cutwaters both sides bearing pedestrian refuges. The bridge is now bypassed by a new bridge not far to the south.

## LLANGOLLEN   SJ 215420   On the north side of the town of Llangollen

The existing bridge over the River Dee with four pointed arches of unequal width is of c1500, but tradition has it that Bishop Trevor of St Asaph built a bridge here in the mid 14th century to replace a 13th century structure of wood. The bridge was widened on the upstream western side in 1873 and widened again in 1968-9. An extra arch for the railway at the northern end was added in 1863. See picture on page 73.

## LUDFORD   SO 512742   On the south side of Ludlow

From the centre of the town Ludlow Broad Street leads southwards under a late 13th century twin-towered gateway in the town wall down towards a mid 15th century bridge over the River Teme with three segmental arches with plain ribs. There are very large cutwaters, one of which once supported a chapel of St Catherine and the others spacious refuges. At the SE corner a squinch arch allows a widening of the road onto the southern arch. The middle arch was repaired after a partial collapse in 2012. Here in 1459 was fought a skirmish between the Duke of York's forces holding Ludlow castle and the Lancastrians. Dinham Bridge further upstream below the castle was of wood on stone piers until the 1820s. See plan on page 17.

*The bridge over the River Dee at Chester*

*Leckwith Bridge, near Cardiff*

*Ludford Bridge over the River Teme near Ludlow, Shropshire*

## LUGG   SO 532518   3km NE of Hereford

The three arches of the bridge carrying A4103 from Hereford to Worcester are all different. Widening on the downstream side to the south in the 1960s has doubled the original width. The pointed eastern arch with ribs is 15th century and may be associated with indulgences granted in 1464 to those who would contribute to repairs. The round central arch, also ribbed, is a later rebuilding using old materials. The narrow four-centered western arch with crudely joggled voussoirs may be 16th century. A 14th century mill once lay just downstream, replacing one upstream mentioned in Domesday Book.

## MERTHYR MAWR   ST 891784   2km SW of Bridgend

The New Inn Bridge is a late medieval structure of four plain pointed arches over the Ogmore River. Of later date are the parapets, with holes to allow sheep dipping on the downstream western side, and the stepped tops of the cutwaters.

## MISKIN   ST 046806   2km south of Llantrisant, 15km NW of Cardiff

The two pointed arches over the River Ely are of 16th century origin but the presence of bricks in the soffits suggests considerable 17th or 18th century rebuilding.

## MONMOUTH   SO 505125 & 502123   On the SW side of the town of Monmouth

The Wye and the Monnow together almost enclosed the walled town. A suburb extended SW to where the Monnow is crossed by a late 13th century bridge which is the only one in Britain to retain a fortified gateway upon it. There are three ribbed segmental arches separated by piers with cutwaters, one of which supports the gateway. It measures 9m by 3.6m and has D-shaped ends pierced since the 19th century by narrow pedestrian walkways flanking a passage 3.5m wide still used for vehicular traffic. The three machicolations on the outer face are a later addition and the corbels between them would have obstructed the raising of the portcullis. The two tiny upper rooms were reached by a spiral staircase on the north side. The wallheads and roof are post-medieval. In 1988 part of an earlier timber bridge was found in the river bank below and dated to the 1170s by tree-ring analysis. To the west is the 16th century Clawdd Du bridge over a ditch protecting Over Monnow. The five-arch bridge over the Wye is also of medieval origin but was rebuilt in 1617 and widened on both sides in 1879. See p9.

## MORDIFORD   SO 569374   6km ESE of Hereford

The round eastern arch, the massive cutwaters and the arches of the causeway to the west are 16th century, when the bridge was slightly widened on the south side. The pointed western arch is 14th century. The corbels below both arches are thought to have been to support diagonal struts to carry an earlier timber bridge at a lower level.

## MORETON   SO 512459   6km north of Hereford

The bridge is set where a tributary joins the River Lugg on the north side. The two eastern round arches are probably 16th century. The segmental west arch is more recent.

## PONT SPYDDYR   SN 434059   9km NW of Llanelli

The Morgans of Muddlescwm, a house to the west of the bridge, are thought to have sponsored this five-arch 15th century structure. It is essentially a causeway with cutwaters with refuges since only one arch actually spans the Gwendraeth Fawr.

## PONT-Y-CLOCHYDD   SO 990119   20km west of Welshpool

The single arch spanning 8m and rising 5m over Afon Twrch is said to have been used by Owain Glyndwr when in transit from Machynlleth to Sycharth.

*The bridge over the River Ogmore at Merthyr Mawr, near Bridgend*

## RHUDDLAN   SJ 022780   On west side of Rhuddlan to the SE of Rhyl

Both the wooden bridge of 1277 and the stone bridge of c1358 must have had a draw-bridge to allow ships on the canalised River Clwyd to reach a dock beside Edward I's fine new castle. It seems there were originally three spans but in the rebuilding of 1595 (the date that appears on a panel on the SE cutwater) two spans (one presumably a drawbridge) were replaced by one large arch. The other arch may have been rebuilt at the same time leaving one massive pier as the sole remnant of the medieval bridge.

*Mordiford Bridge over the River Wye near Hereford*

## SHREWSBURY   SJ 488128 & 496123   On either side of the town centre

The medieval walled town lay within a loop of the River Severn and had gates forming part of two bridges. The Mardol Gate with two round turrets on the NW side lay on a pier of what was originally called the St George's Bridge, but which since at least the 1330s has been known as the Welsh Bridge. In 1575 the drawbridge in front was replaced and the upper part of the SE side rebuilt (see page 3). The Welsh Gate lay at the far end of the bridge, the upper room of which remained in use as a guardroom until it was demolished in 1773, whilst most of the bridge was replaced in 1795. One surviving flood arch and the base of the twin towers of the Welsh Gate were excavated in 2006. The towers replaced an earlier pair of turrets which itself replaced a 12th century tower.

Floods in 1542 caused the collapse of the Stone Gate facing the English Bridge. The bridge of six arches and causeway with a further eleven arches extended 290m from the gate to the Benedictine abbey, there originally being two wide but shallow channels of the river here with a central island. The main bridge was replaced in 1769-75. Sections of the causeway are thought to remain underground as solid walls have been found occasionally during works on roads and buildings.

## TENBURY WELLS   SO 596686   10km SE of Ludlow

The three northern arches of the bridge over the River Teme forming the boundary between Worcestershire and Shropshire are medieval. They each have four ribs of sandstone. The bridge was widened both sides in 1908 with a concrete top deck. See p15.

## OTHER BRIDGES PROBABLY EARLIER THAN c1600 (not marked on map).

AFON TRWSGL SH 546493 Four (was five) clapper slabs set on drystone abutments.
BEDDGELERT  SH 590481 Medieval in origin but mostly rebuilt c1625,1778 & 1802.
BRECON  SO 043286 Seven segmental arches of 1563 over River Usk. Widened.
CAERGWRLE SJ 305576 Four-arch pack-horse bridge over River Alun restored 2000.
DOL-Y-MOCH  SH 685416 Four segmental arches with refuges over Afon Dwyryd.
GOODRICH  SO 577200  Fine multi-chamfered arch of c1300 near castle entrance.
LLANGYNIDYR SO 152203  Slightly curved six-arch bridge of c1600 over River Usk.
LLANSTINAN SM 945332 Seven lintels set on drystone abutments. Upper deck 1988.
LOWER HUXLEY HALL SJ 498623 Two segmental arches with cutwaters over moat.
PENMACHNO  SH 806529  Single round-arch packhorse bridge over a small gorge.
PONT HELYGOG SH791196 Remains of wider ancient bridge below existing bridge.
PONT SARN DHU  SH 711516  Eight-span clapper bridge. Northern piers rebuilt.
PONT-Y-PORTHMYN SH 803398 Single depressed pointed arch over Afon Taihirion.
PORTHCLAIS SM 741242 3m long clapper slab across the Alun River.
RISBURY  SO 540549  Two low round arches over Humber Brook near later bridge.
ROMAN BRIDGE SH 648527  Three span clapper bridge probably of c1520.
RUSHBURY SO 513915  Segmental arch without parapets over Eaton Brook.
ST DAVID'S SM 750254 Arch over leat to SW of cathedral close may be medieval.
STRETTON  SO 474434  Much widened medieval single arch carrying a farm track.
TALGARTH  SO 155337  Segmental arch over River Ennig later widened both sides.
TAN-Y-FYNWENT SH 330816  Clapper bridge associated with medieval corn mill.
WILTON  SO 590242 Six round arches of 1597, ribs and large cutwaters. Widened.
YSGOLDY  SH 634514  Clapper bridge of six spans below SW end of Llyn Gwynant
OTHER PLACES FORMERLY WITH MEDIEVAL STONE BRIDGES:
  Atcham, Carmarthen, Ewenny Priory, Haverfordwest, Llandeilo, Llandysul,
  Llawhaden, Montford and Welshpool. Bridges at Buttington, Caerleon, Cardigan,
  Llanidloes, Newport and Newtown were only of wood during the medieval period.

*Bridge over the River Dee at Llangollen*

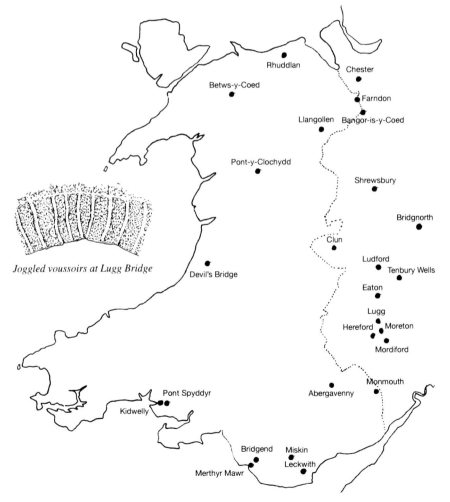

*Joggled voussoirs at Lugg Bridge*

SURVIVING PRE-REFORMATION BRIDGES IN WALES AND THE MARCHES

# MEDIEVAL BRIDGES OF NORTHERN ENGLAND

## ALDBROUGH  NZ 202114   12km  NE of Richmond

This packhorse bridge over the Aldborough Beck has two pointed arches and a big pier dividing them from a flatter and probably later arch to the south. The roadway is only 1.3m wide between parapets added later on.

## ALDIN GRANGE   NZ 249429  3km WNW of Durham

The monks of Durham cathedral-priory built this three-centered arch on simple imposts spanning 13.5m over the River Browney.  The arch has two chamfered arch-rings.

## AUSTWICK    SD 774686 & 768681   7km NW of Settle

Flascoe bridge has five clappers on four piers across the Austwick Beck. To the SW is the Pant Bridge with a flagged causeway linking a bridge of three clappers with a single clapper to the NW.  Both are said to be 15th century. See picture on page 5.

## AYSGARTH   SE 011886   40km west of Northallerton

The high single arch bridge spanning 18m over the River Ure just above Aysgarth Falls is said to be of 1539. It has a recessed inner arch-ring. The bridge was doubled in width in 1788 on the downstream eastern side.

## BALDER  NZ 009200    Near Cotherstone  7km  NW of Barnard Castle

The single pointed arch over the River Balder close to its confluence with the Tees is medieval in origin but rebuilt in the 1680s and later widened on the upstream side with two extra ribs without the chamfers seen on the older four.

## BARFORTH or BARTON   NZ 163161   1km east of Barnard Castle

This is a widened single-arch 14th century bridge over Chapel Gill, near a ruined chapel of St Lawrence. There are three chamfered ribs and parapets carried on corbelling.

## BARNARD CASTLE   NZ 048164   Below the west side of Barnard Castle.

Commanded by the castle, this bridge over the River Tees has two pointed arches each with four stepped arch-rings and four ribs. The ends of the refuges are solid, although this is not an ancient arrangement since until c1800 a chapel stood on one of them. That proves that the pier is medieval even though the arches either side of it may per-haps be a rebuild of 1596, the year that appears on a stone reset at the western end.

## BAYSDALE   NZ 621067   18km south of Redcar

This 13th century single arch over the Black Beck was built by the nearby nunnery. The arch has four plain ribs.

*Bowbridge in Wensleydale*

*Aysgarth Bridge in Wensleydale*

*Bow Bridge near Furness Abbey*        *Croft Bridge on the River Tees. The right-hand ribs are later*

## BOROUGHBRIDGE    SE 396670    At Boroughbridge, 22km NW of York

There was only a wooden bridge over the River Ure here in 1322 when Thomas, Earl of Lancaster was defeated and captured nearby, but Leland in the 1530s saw the existing bridge of sandstone ashlar. A segmental arch is flanked on either side by a pointed one, each having five ribs. It was widened on the upstream western side in 1785 and widened again in 1969. There is no easy access to the sides of the bridge.

## BOW    SD 223714    3km NE of Barrow-in-Furness

This packhorse bridge over Grange Beck, perhaps of 15th century date and built by the monks of nearby Furness Abbey, has three low round arches over the Mill Beck.

## BOWBRIDGE    SD 934910    Near Askrigg, 19km west of Leyburn

This bridge over the Scargill Beck (a tributary of the River Ure) now lies in a field and has been extended southwards to give a width of about 7m. The original semicircular 13th century arch at the north end has four ribs and a chamfered arch-ring.

## CATTERICK    SE 227994    5km SE of Richmond, 15km NW of Northallerton

Widening and refacing in 1792 mostly hides what remains of the bridge over the River Swale built in 1422-5, with William de Burgh as patron. The contract document mentions £170 for masons' labour. In 1505 a chantry chapel was built on the SE side of the bridge. However the bridge seems to have been mostly rebuilt c1565-90 with the middle two of four arches being four-centered. Originally there were just three arches.

## CHESTER NEW BRIDGE    NZ 284523    1.5km NE of Chester-le-Street

A driveway up to Lambton Castle crosses the River Wear by means of the 15th century Chester New Bridge of four arches with cutwaters with battered plinths. The arches each have five chamfered ribs and have chamfers on the two outermost of three arch-rings. The parapets are 19th century. This bridge carried a main road until a new bridge was built further west in 1924-6. See picture on page 77.

## CROFT    NZ 290099    4km south of Darlington

This bridge over the River Tees, here the boundary between County Durham and North Yorkshire, is a 15th century structure of seven pointed and ribbed arches built of red sandstone. The sixth arch is lower and the seventh arch is smaller and semi-circular. The bridge was restored in 1673 and in 1795 it was doubled in width on the upstream NW side, giving each arch a total of 11 chamfered ribs of which five are original. On the downstream side the parapet is corbelled. Both sides have polygonal refuges.

## COVERHAM   SE 046985   5km WSW of Leyburn

A single pointed late medieval arch springs from chamfered imposts and spans 15m over the River Cover. Built by the nearby abbey, it was recorded as decayed in 1615.

## DEEPDALE   NZ 045166  At Startforth, just NW of Barnard Castle

This bridge over a beck joining the Tees nearby was recorded as ruinous in 1605. Widening of the ribbed segmental arch on the eastern downstream side out to a new face with three slightly recessed arch-rings probably happened shortly after then. Widening on the other side in the 18th century doubled the width.

## DILSTON   NY 976634   3km east of Hexham

Part of an abutment and one pier each with the springing of chamfered ribs remain of a 14th century bridge over the Devil's Water.

## DURHAM   NZ 272425 & 275424   On either side of Durham city centre

The original Norman walled town enclosing the cathedral lies on a steep-sided peninsula almost surrounded by a loop of the River Wear. The two ancient bridges connect with the market place part of the city to the north of the castle, which was not walled until the 14th century. The Framwellgate bridge on the west was built by Bishop Flambard c1128. One arch possibly of that period remains under buildings at the east end, where there was a gatehouse. It is possible that shuttering used to build the vault over the nave of the cathedral was later used to construct this arch. The existing bridge of two wide elliptical arches of three orders set upon closely-grouped ribs was built after flood damage in 1401. Chamfering appears only on the outer edge of the outer ribs. Some of the ribs look as if they have been renewed. The bridge originally had a chapel in the middle and was widened on the northern upstream side in 1856. See page 19.

In the 1170s Bishop Hugh le Puiset rebuilt the Elvet bridge on the east side. The existing structure with arches of three orders with five ribs is late 13th or early 14th century. There are five arches crossing the river and several land arches, those at the western end obscured by buildings, whilst the first land-arch on the east is semi-circular and could be 12th century, although it has been much widened northwards. The bridge has cutwaters but no pedestrian refuges. Leland's description of c1540 suggests there may have originally been fourteen arches in total. By the late 16th century there were houses upon this bridge. Three arches were rebuilt after being destroyed by flooding in 1771, which explains why some of the ribs look new, and the bridge was widened on the northern or downstream side in 1804-5. At the west end a House of Correction of 1632 with two cells below a land arch has replaced a chapel of St James. The Dutch-gabled shop projecting over a cutwater at the east end has replaced a chapel of St Andrew which existed by c1218-34.

The Prebends' Bridge was built after the floods of 1771 but on the west side an abutment of the previous 17th century bridge is still visible.

## EAMONT   NY 523288   At Eamont Bridge, 1.5km SE of Penrith

A bridge of three segmental arches with cutwaters crosses the River Eamont which once formed the boundary between Penrith in Cumberland and Brougham in Westmorland. Each arch has four original plain ribs and two more under widening on the downstream western side. There are cutwaters on each side but only the larger ones facing upstream to the east carry refuges, which are semi-hexagonal in plan. An indulgence of the 1420s helped provide funds for building this bridge.

*Chester New Bridge over the River Wear in Co Durham*

*Elvet Bridge at Durham*

*Eamont Bridge near Penrith*

## EDISFORD or EADSFORD    SD 726415    2km SW of Clitheroe

Three of the four arches across the River Ribble may be 16th century, but the central arch twice the width of the others must be later. Corbels for shuttering the arches survive on the piers, which have cutwaters. The bridge was widened in the 19th century and has been much rebuilt. There are five more arches in the causeway to the west.

## FELTON    NU 185003    14km north of Morpeth

This bridge carrying the old Great North Road over the River Coquet is probably 15th century. It has three ribbed arches and semi-hexagonal pedestrian refuges. It was widened on the upstream western side in the late 18th or early 19th century. There is a skew arch across the NE corner on the downstream side,

## FORD    NT 939375    10km NW of Wooler

One 16th century arch with four plain ribs bearing masons' marks still survives. The other arch, which was wider and humped, was rebuilt in 1807-8 after collapsing whilst being widened on the upstream side.

## FOUNTAINS ABBEY    SE 273682    National Trust property, 5km SW of Ripon

Several ranges of the abbey buildings straddle the River Skell. A 13th century bridge of three round ribbed arches gives access to the guest apartments from the south. To the west is a fine late 14th or 15th century bridge of two ribbed arches with cutwaters providing access between the precinct and the abbey mill south of the river. See p11.

## GILLING    NZ 183052    4km north of Richmond

The three segmental arches over the Gilling or Skeeby Beck are of 1799, when the bridge was considerably widened. A flood arch to the south has five chamfered ribs and must be a relic of the bridge mentioned in the 15th century.

## HARTFORD    NZ 242800    7km SE of Morpeth

This bridge over the Blyth was mostly rebuilt in 1904 and had already been widened on the west side in 1688, but the triple-chamfered northern arch is late medieval.

## HEBDEN    SD 992273    At Hebden Bridge, 11km WNW of Halifax

The bridge of c1510 across Hebden Water was repaired in 1600, 1602 and 1657. The parapet was repaired in 1845 and raised in 1890. Cutwaters divide two segmental arches with slight chamfers over the river and a third over land at the east end. Three of the cutwaters rise up into refuges.

## HELMSLEY    SE 614835    South of Helmsley, at SW corner of North York Moors

The bridge over the River Rye was mostly rebuilt in the late 18th century but the slightly pointed main arch and a narrower, more sharply second pointed arch to the north preserve medieval forms. The bridge has been widened on the western upstream side where a cutwater is rounded above a triangular base. There is no access to the sides.

## HODDER    SD 704391    5km SW of Clitheroe, 11km north of Blackburn

The River Hodder, once the county boundary between Lancashire and Yorkshire, is crossed by a ruined medieval packhorse bridge of three segmental arches without parapets. It is mentioned in 1430 but the central arch at least, now reduced to a bare single arch-ring, is a rebuild of 1561. In 1648 Cromwell's army passed over this bridge on their way to gain a victory over the Royalists at Preston.

*The bridge from which Hebden Bridge takes its name*

*Packhorse bridge at Hodder with the middle arch reduced to a single ring*

*Plain ribs under an arch at Ford*          *Kirkham Bridge: only the nearest of the three arches is medieval*

*Kirkby Lonsdale Bridge over the River Lune*

*Kildwick Bridge*

## HOWDEN   SE 351920

2km SW of Northallerton

Now bypassed by the A684, this 19th century bridge over the River Wiske has two pointed eastern flood arches with chamfered ribs which are medieval.

## INGHEY   SD 961517   3km west of Skipton

Flood arches at either end are thought to be 13th century. The main bridge over the River Aire is of 1773, but incorporating one mid 17th century arch at the east end. It was widened on the south side in the mid 19th century.

## KILDWICK   SE 011457   7km NW of Keighley

Built by Bolton Priory, and bypassed since 1988, this 14th century bridge over the River Aire has two pointed arches and two segmental ones, each with ribs and cutwaters. It was widened on the east side in 1780 and given new parapets in the 19th century.

## KILGRAM   SE 191860   19km NW of Ripon

John Leland c1540 referred to this bridge over the River Ure to the east of Jervaulx Abbey as "the great old bridge". There are six segmental arches each with four plain ribs and a slight chamfer on the arch-ring. The end arches are smaller and lower.

## KIRKBY LONSDALE   SD 615783   SE of Kirby Lonsdale, 20km NE of Lancaster

Built by St Mary's Abbey at York, the fine 15th century Devil's Bridge over the River Lune has three semi-circular triple-chamfered arches each with four ribs and divided by semi-hexagonal refuges. The arches span 16.7m and rise 12m above the river. There is a sundial dated 1673 on the south side. Since 1932 the bridge has been bypassed.

## KIRKHAM   SE 733658   8km SW of Malton

The pointed northern arch of this bridge over the River Derwent just below Kirkham Priory survived a rebuilding of 1806 probably because it was the responsibility of the North Riding of Yorkshire, whilst the other arches were maintained by the East Riding.

## KNARESBOROUGH   SE 345572   West of Knaresborough, 5km NE of Harrogate

The two original medieval arches of the High Bridge carrying A59 over the River Nidd each have four chamfered ribs. The arches were widened on either side in 1826, and there was a further widening of the upstream western side in 1929.

*The bridge over the River Nidd at Knaresborough*

## LESBURY     NU 238116     5km ESE of Alnwick

Cutwaters carried up as refuges divide a pair of triple-chamfered arches of this 15th century bridge over the River Aln. The northern arch is segmental and the other is pointed, each spanning 10m. The bridge was doubled in width on the downstream side.

## LOYN   SD 582697   NNW of Hornby,  9km east of Carnforth

Three late medieval segmental arches with recessed inner arch-rings cross the River Lune below the earthworks of Castle Stede. There are refuges between the arches.

## MARSKE     NZ 104004     7km west of Richmond

The segmental arch with chamfered ribs and a hoodmould over the Marske Beck is 15th century, but repaired in 1588. It was widened on the NW upstream side in the 19th century and the corbelled parapets above the abutments may date from then.

## MORPETH     NZ 200859      In the middle of Morpeth

In 1869 an iron footbridge was built over the surviving central pier and abutments of the medieval bridge across the River Wansbeck. The rest had been blown up in 1834. By the north end is a 13th century chantry chapel later adapted as the parish church of All Saints. The south and east parts of the chapel were remodelled in 1738 when a new arcade was provided inside. Latterly it has served as a tourist information office.

*At Morpeth in Northunmberland the medieval bridge is now reduced to just the central pier*

*Old print of the castle and the bridge with its gatehouse at Newcastle-upon-Tyne*

## NEWCASTLE   NZ 253637   On the SE side of Newcastle-upon-Tyne

A timber bridge was destroyed by fire in 1248. One arch hidden under the northern abutment of the Swing Bridge is all that remains of a late 13th century bridge over the River Tyne which lasted until flooding in 1771 destroyed three arches, killing six people. A prison and chapel lay on the ten-arch bridge. The southern four arches were the responsibility of the bishops of Durham, whose arms lay on a tower at the south end.

## NEWTON CAP   NZ 205304   1km NW of Bishop Auckland

A large pier with almost rectangular refuges on the cutwaters carries two very long arches over the River Wear which may go back to the time of Bishop Walter Skirlaw, 1388-1406, although they could be as late as the 16th century. The slightly pointed southern arch spans 28m, and the segmental northern arch spans 30m, most of it usually dry land. Each arch has three arch-rings with the lower ones recessed. See p17.

## OTLEY   SE 201459   On NW side of Otley, 16km NW of Leeds

Five segmental arches each with four ribs remain of the bridge of 1228-9, only the outer edges of the outer ribs being chamfered. The river bed is paved below the bridge. In 1775-6 the bridge was widened on the upstream western side and provided with an extra flood arch at each end. The cantilevered footways are a modification of 1957.

## PAYTHORNE   SD 831513   3km north of Gisburn,   13km west of Skipton

There are three ribs under the segmental arch over the River Ribble. The bridge has been widened on the upstream northern side and two flood arches have been added.

*The spectacular Newton Cap Bridge near Bishop Auckland*

*Bridge near Prudhoe Castle, Northumberland*          *Pickering Bridge, North Yorkshire Moors*

## PICKERING    SE 796841    In Pickering town centre,  12km north of Malton

One late medieval flood arch survives with three widely spaced chamfered ribs. The arches crossing the river are 18th century. The south side has been widened later on.

## PIERCEBRIDGE    NZ 214155 & 211155   8km west of Darlington

The village lies on the site of a Roman fort of c260-70. A substantial civilian settlement to the east was traversed by Dere Street which originally crossed the River Tees 370m east of the fort. The south abutment two courses high and footings of four of the ten piers still remain of a later Roman bridge built further east. The large stones were fastened by iron clamps and timber decking carried the roadway whilst the river bed was paved to reduce scouring of the piers. Mortar only appears to have been used in the later causeway built at the south end after silting resulted in a widening of the river.

In the 13th century the Roman bridge was replaced by a new one further west. It survived in use until replaced c1500 by the existing bridge of five arches which was seen by John Leland c1540 and greatly widened in 1781. The arches have three arch-rings with the inner ones recessed. See photo of Roman bridge on page 5.

## PRUDHOE    NZ 102634    15km west of Newcastle-upon-Tyne

Just east of the unwalled outer enclosure of the castle is a bridge over a burn with a pointed outer arch facing south and semi-circular ribs further in, whilst the northern side has been slightly widened.

## RIEVAULX    SE 574843    4km west of Helmsley, 34km north of York

Extensions of the 18th century mask a medieval bridge over the River Rye. Two of the three arches may go back  to the 13th century, when a bridge is recorded here.

## RIPON    SE 317720    1km NNE of Ripon

The North bridge over the River Ure mentioned in 1309 has seven river arches, some segmental and others pointed, and five more arches in a causeway to the SW. The bridge has been widened twice on the upstream side and has been much rebuilt. One arch has a row of trefoils below the parapet. See page 84.

*The North Bridge at Ripon*                    *Bridge Chapel at Rotherham*

## ROCHDALE    SD 896133    NNE of Rochdale town hall

Early in the 20th century seven bridges over the River Roch were joined together to make a culvert 500m long. Rediscovered in 1994, one of the bridges is 14th century and is in the process of being re-exposed as part of a town centre regeneration scheme.

## ROTHBURY    NU 059016    On south side of Rothbury, 20km NW of Morpeth

The three northern arches of this bridge over the River Coquet are medieval and have chamfered ribs. The southern arch is 16th or 17th century. The bridge was widened on the eastern downstream side in 1759 and is said to bear that date low down with the name William Oliphant (a local mason).  A new concrete and steel superstructure was provided in the mid 20th century. Further work was done on the bridge in 2012.

## ROTHERHAM    SK 428931    On west side of Rotherham town centre

This bridge with four pointed arches with ribs over the River Don was widened in the 18th century but reduced to its original width in the 20th century, a rare instance of such a sequence of events. It is, however, truncated at the western end. Money was willed by John Bokyng in 1483 towards the building of the bridge chapel. It has pairs of renewed three-light windows on each side and a renewed north (ritual east) window of four lights. The plain south facade towards the bridge has a two-light window set over a doorway. The chapel became an almshouse in 1569 and in 1779 it became a prison, with a deputy constable residing in the former chapel itself and the prisoners being held in the tunnel-vaulted crypt below. The chapel later served as a shop. It was only restored as a chapel in 1924.

## SALTERS    NZ 254685    1km east, Gosforth, 5km north of Newcastle-upon-Tyne

This bridge over the Ouse Burn has a round western arch and a pointed eastern arch with a cutwater between them. It has been widened.

## SESSAY    SE 464747    26km NW of York

The arch over the Birdforth Beck to the SW of Sessay Church appears to be medieval and has five chamfered ribs, but the bridge has been much rebuilt. It bears a shield with arms of the Dawnay and Pagot families.

## SETTLE    SD 816641    NW of Settle,   32km east of Lancaster

The bridge over the River Ribble has a pair of four-centered arches each with four chamfered ribs. It was widened on the downstream southern side in the 1790s.

*Sunderland Bridge over the River Wear*

*The tiny arch-ring over a dry stream-bed on the village green at Sinnington*

*Twizel Bridge in Northumberland*

## SHERBURNHOUSE   NZ 306416   3km ESE Durham

This delapidated arch with four ribs spanning 5m over the Sherburnhouse Beck carried what is now the A181 until the 1930s.

## SINNINGTON   SE 744858   Below North York Moors, 5km WNW of Pickering

There is a tiny medieval bridge no longer crossing any stream in the middle of the village green. Nearby is a bridge of 1767 across the River Seven.

## SOWERBY   SE 059235   At Sowerby Bridge, 4km SW of Halifax

The bridge of three round arches dating from 1514 over the River Calder was widened in 1632 and again in 1875.

## STANHOPE   NY 985390   At Stanhope, 23km NW of Bishop Auckland

A round arch of 11m span crosses the River Wear and has three chamfered ribs. It was doubled in width in 1792 and provided with new parapets in 1837.

## SUNDERLAND   NZ 264378   5km south of Durham

The middle two semi-circular arches of this bridge over the River Wear are 14th century. Each has five shallow ribs, the only chamfering being on the outer edges of the outermost ribs. The outer two arches were rebuilt in 1770. One of the cutwaters with large pedestrian refuges has been cut back to a square at the top. The northern approach to the bridge was rebuilt following an accident in 1821.

## TILEHOUSE   SE 680850   2km SW of Kirbymoorside, 7km east of Helmsley

The single sandstone arch with a moulded band above it which crosses the Hodge Beck south of the A170 is thought to be 13th century.

## TWIZEL   NT 885434   5km NE of Coldstream

The gorge of the River Till is crossed here by a late medieval single arch supported on five closely spaced chamfered ribs and having four recessed arch-rings. It rises high above the river and spans 27m. The corbelled parapet appears to be later. See p85.

*Wensley Bridge over the River Ure*

*Gatehouse at Warkworth*

*Warkworth Bridge*

## WAKEFIELD    SE 337201    To the SE of the centre of Wakefield

This bridge with nine pointed arches existed by 1342, when toll rights were granted. The arches have two arch-rings, both of them chamfered and the northern ones have ribs. There has been widening on the upstream west side. Projecting from the middle of the east side is a bridge chapel which was licensed in 1357. It is an unusually ornate structure with a five-light end window flanked by a corner staircase turret, and three square-headed windows of three lights on each side, above which are hoodmoulds on head-stops and a parapet with blank panelling. The staircase leads down to a sacristy below the eastern part of the chapel. The west facade of five tiny bays with three doorways towards the bridge was entirely replaced in 1847, and was again renewed in 1939-40. Most of the rest of the superstructure was rebuilt in 1847 with old materials. As early as c1580 Camden had described it as much defaced. The original west front survives reset on a boathouse at nearby Kettlethorpe Hall. See photo on page 11.

## WARCOP    NY 748155    8km SE of Appleby

This is the only surviving ancient bridge over the River Eden and is probably a 16th century successor to the bridge mentioned in 1374 and 1380. It has three red sandstone segmental arches each with four plain ribs and semi-hexagonal refuges.

## WARKWORTH    NU 248062    9km SE of Alnwick

The small town is almost surrounded by a loop of the River Coquet with the castle guarding the only landward approach on the south side. At the northern end is a late 14th century bridge with two ribbed arches each of 18m span with between them a cut-water carried up as a pedestrian refuge. John Cook of Newcastle left 20 marks towards construction work in 1379. At the south end is a plain ruined gatehouse measuring 8.3m by 5.5m. An upper room is reached by a staircase in the east wall. The vaulted gateway passage is flanked on the other side by a tiny guard room.

## WENSLEY    SE 091894    2km SW of Leyburn, 27km NW of Ripon

Of the four arches over the River Ure the two that are still pointed probably date from the early 15th century since Lord Scrope's will of 1400 left money towards works on this bridge. The inner arch-rings have a slight chamfer and the outer ones a bolder chamfer. The south arch is later. The widened upstream side of c1820 has segmental arches.

## WESTERDALE    NZ 663061    15km south of Saltburn-by-the-Sea

This small medieval bridge known as Hunters Sty, with a ribbed arch over the River Esk, carries just a footpath. It may be as old as the 13th century. The parapets were restored in 1874.

**WETHERBY**   SE 404480   On south side of Wetherby, 12km SE of Harrogate

Two of the six arches of the bridge over the River Wharfe are probably medieval. It has been widened on both sides. Close to it are slight remains of a castle. This bridge carried the Great North Road towards Scotland coming up via Wansford, Stamford, Newark and Ferrybridge. Originally just 3.5m wide, it was widened in 1773 and 1826.

**WHALLEY**  SD 732359        South side of Whalley, 10km NNW of Blackburn

Ribs remain under the middle and northern arches of the 14th century Abbey bridge crossing the River Calder in 10m spans. The bridge was twice widened on the down-stream side before 1914, when it was widened again, this time on both sides.

**YARM**   NZ 418131   North of Yarm, 7km SSW of Stockton-on-Tees

The "bridge of Yarum" mentioned in 1228 was probably of wood. The three middle arches, pointed with two chamfered arch-rings, probably represent what Bishop Skirlaw had built c1400. One arch has ribs. There was a drawbridge instead of a stone arch at the north end during the Civil War. The existing end arches are round and may date from 1799, when the bridge was widened on the downstream east side.

## OTHER BRIDGES PROBABLY OLDER THAN c1600 (not on map opposite)

BAINBRIDGE SD 935901 Single pointed 16th century arch widened in 1785.
BELLASIS NZ 190776  Two segmental arches over River Blyth may be 16th century.
CLOW BECK NZ 280100  Two arches set at angle. String course over west arch.
COW SD 827569  Probably an 18th century rebuild, although the arches are ribbed.
CROOK OF LUNE SD 620963  Two segmental arches over the River Lune.
EGGLESTONE NY 062152  Packhorse bridge over Thorsgill Beck, NE of abbey ruins.
ELVINGTON SE 705477  Looks medieval but generally considered to be 17th century.
GRINTON SE 046985  Irregular north arch. Segmental south arches later. Widened.
HAMPSTHWAITE  SE 260591   Three segmental arches of 1598, rebuilt in 1640.
HELL GILL SD 787969  Rough single arch of uncertain date over Hell Gill Beck.
HIGHERFORD  SD862401   16th/17th century packhorse bridge over Pendle Water.
HIGH SWEDEN  NY 379067  Five-centered arch over Scandale Beck. No parapets.
IVELET SD 933977  Round arch with hoodmould. Probably late 16th century.
KILLINGHALL  SE 287597 Two segmental ribbed arches, four flood arches. Widened.
LINTON  SD 998628  Packhorse bridge & clapper bridge near village green.
LOW DINSDALE  NZ 346110  Small segmental arch over drained inner moat of house.
NEWBY SD 369863  Five segmental arches over R. Leven, may be late 16th century.
NEWSHOLME DEAN  SE 019405  Cantilevered clapper bridge next to later bridge.
ROSSENDALE SD 811226  Twin segmental arches over R. Irwell. Cutwater to east.
SOUTH CHURCH NZ 216283  Only the abutments are still medieval.
STOKESLEY NZ 524085   Round arch over R. Leven, hoodmould and tall parapets.
STRINES  SD 959285  Segmental arch over Colden Water approached by steps.
SWANSIDE SD 785454   Segmental arch over Smithies Brook.
THORNS GILL SD 777794  Single round arch without parapets over Gale Beck.
WILBERFOSS  SE 732509  Single arch west of the church perhaps 16th century.
WYCOLLER SO 932395  Small clapper-bridge close to small two-arch bridge.

*Westerdale Bridge on the Yorkshire Moors*

*Yarm Bridge over the River Tees*

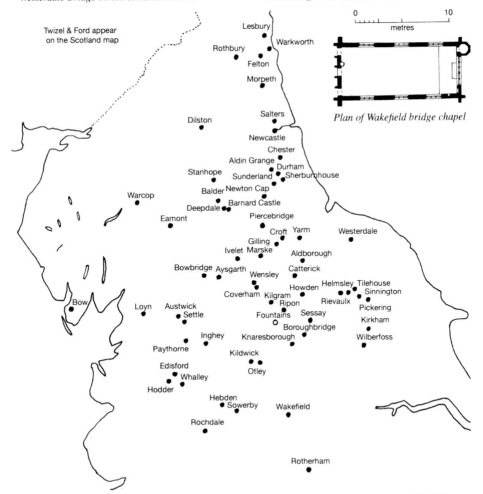

*Plan of Wakefield bridge chapel*

*SURVIVING PRE-REFORMATION BRIDGES IN NORTHERN ENGLAND*

# MEDIEVAL BRIDGES OF SCOTLAND

## ABERDEEN    NJ 928036    3km SW of the centre of Aberdeen

Widened from 4m to 7m in 1841, the Bridge of Dee to the south of the city was built in 1520-7 by Bishop Dunbar from funds accumulated for it by his predecessor, Bishop Elphinstone. There are seven ribbed arches each of 15m span. There are heraldic panels and inscriptions on the piers and a sundial beside a stairway at the south end. The north end originally had a gatehouse and a chapel with a statue of the Virgin Mary. The Ruthrieston bridge not far to the north is a transplanted and rebuilt structure of 1693.

## ALLOWAY    NS 32178    3km south of Ayr

The Auld Brig O' Doon is famous for its association with Robert Burns, and has parapet copings which look like hand-rails, as referred to in the poem Tam O'Shanter. It is said to be the work of Bishop Kennedy, d1466, but could be a rebuild of after 1593, when it was reported to be ruinous. The arch spans 21.5m and has a cobbled roadway.

## ALYTH    NO 245487    At Alyth, 7km NE of Blairgowrie

The packhorse bridge with two segmental arches of unequal size over the Alyth Burn is of 16th century origin but much repaired and altered, the parapets being of c1800.

## AYR    NS 339221    At the north end of Ayr

The two northern arches each spanning 15.5m and the cutwaters could represent what was under construction in the 1490s. The slightly misshapen taller pointed southern arch looks older, although it may be a 17th or 18th century rebuild.

## BALGOWNIE    NJ 941097    3km north of the centre of Aberdeen

This high and wide single arch with a moulded edge rising 17m above the River Don may be the oldest remaining bridge in Scotland. Robert Bruce is said to have completed it in the 1320s although the work may have been begun by Bishop Cheyne in the 1290s. Spanning 12m, this is the finest single-arch pre-Victorian bridge in Scotland.

## BANNOCKBURN    NS 807904    4km SSE of Stirling

The taylor Robert Spittal paid c1530 for the erection of a single arch with a 10m span over the Bannock Burn. It is built of ashlar with chamfered voussoirs and was repaired in 1631 and 1710. It was widened on the western upstream side in 1781.

## BRECHIN    NO 604593    To the south of Brechin, 12km west of Montrose.

The southern of two segmental arches carrying A933 high over the River Esk is thought to be the work of Bishop Crannock c1469. An earlier bridge is said to have been built by Bishop Gregory c1219-46. The northern arch was rebuilt in 1786, when the bridge was widened on the western upstream side and provided with new cutwaters, that on the east side having a shallow refuge on top.

## CRAMOND    NT 179753    8km ENE of Edinburgh

The three arches of this bridge of c1500 are all slightly different as a result of rebuilding of the eastern arches in 1617-19, following flood damage of c1587. Other repairs were made in 1689-91, 1761, 1776 and 1854. A string-course is stepped over the arches, two of which are pointed and the third is nearly rounded. The western arch is double-chamfered and has four chamfered ribs which are bridged by slabs. The drain holes are now a full metre below road-level. Here in 1535 Jock Howieson, armed just with an agricultural flail, aided James V when alone in disguise and assailed by a band of gypsies, and was rewarded by a grant of the king's farm of Braehead. Another dramatic scene occurred here in 1565, when the Earl of Bothwell took possession of Queen Mary.

*Bridge over the River Dee at Aberdeen*

*Brechin Bridge*

*The SW arch of the bridge at Ayr*

*Cramond Bridge*

*An abutment of the bridge at East Linton*

## DAIRSIE   NO 415161

10km west of St Andrews

The east side of this bridge over the River Eden bears a panel with arms of James Beaton, Archbishop of St Andrews, who held the see from 1522 to 1538, but a bridge here is mentioned in 1496. A pier with cutwaters on both sides divides a northern arch with four ribs from two other arches, divided by a pier with a cutwater only on the western upstream side. Upon that pier is a refuge of the 18th century, when new parapets were provided. The roadway over the bridge rises from south to north.

## DOUNE   NN 722012   10km NW of Stirling

An heraldic panel on the parapet of the Bridge of Teith to the SW of Doune records that it was commissioned by Robert Spittal in 1535. There are two wide semi-circular arches with a faceted cutwater. The bridge was widened on the west side in 1866.

## DUMFRIES   NX 969761   In the middle of Dumfries

Devorguilla, Countess of Galloway is said to have erected a bridge here in the late 13th century. The slightly humpbacked Old Bridge to the south of the New Bridge of 1791-04 (widened in 1892-3), was built c1430-2 and the obtusely pointed western arch may be of that period. The other arches of rough ashlar are almost round-headed and may be mostly of 1620 following flood damage, although irregularities suggest more than one period is represented. Only the central cutwaters rise up into refuges, the others having sloping tops. Weepholes allow water from the roadway to drain away through the parapets. Reclamation of land on the east bank of the River Nith allowed the bridge to be shortened from nine arches to six in the early 19th century.

## EAST LINTON   NT 592772   9km ENE of Haddington

The two sandstone arches with closely-set ribs were rebuilt after destruction by the French in 1549. The cutwaters project sufficiently from the large pier set on a rock to carry widening outer arches probably of 1763 on each side.

## GUARDBRIDGE   NO 452189 & 450199   6km NW of St Andrews

Now bypassed close to the south by a bridge of 1935-7 is a bridge of six arches over the River Eden bearing on the north side the arms and initials of James Beaton, Archbishop of St Andrews 1522-39. Bishop Henry Wardlaw is credited as building the bridge in 1419. The wide but shallow refuges on the cutwaters probably date from one of the repair campaigns recorded c1601, 1678-86, 1786 and 1802. The two eastern arches may be of c1530. Also superseded by a modern bridge alongside it is the Inner Bridge over the Motray Water 0.7km to the north. A structure of three semi-circular arches with cutwaters facing upstream to the west, it was repaired in 1598 and widened on the downstream eastern side in the 18th century.

*The Auld Brig o'Doon at Alloway*

*Abutment at Dairsie Bridge*

*Upper North Water Bridge*

*The bridge of Balgownie, near Aberdeen*

*The old bridge over the River Nith at Dumfries*

*The Abbey Bridge near Haddington*

## HADDINGTON    NT 533746 & 519738    On the east side of Haddington

Two bridges each of three arches over the River Tyne were both associated with the former Cistercian nunnery. The Abbey Bridge 1.5km east of the town is 15th century and has cutwaters with refuges, the parapet on the west side being carried on a corbel table. Each arch had five chamfered ribs but the outermost arches have each lost three of their ribs. The smaller and plainer Nungate bridge closer to the town centre is 16th century. The extra arches on the east bank have replaced a former steep approach.

## JEDBURGH    NT 651203    On the south side of the town of Jedburgh

Now used only by pedestrians, the 16th century Canongate Bridge over the River Jed is hump-backed with three segmental arches. A gateway in the middle was repaired in 1677 but removed in the late 18th century. The ribbed western arch now only spans a footpath replacing a former mill lade. Triangular cutwaters support semi-hexagonal refuges, the eastern ones having chamfered rebates possibly from the Franciscan friary.

## MUSSELBURGH    NY 341725    8km east of Edinburgh

Three segmental arches with cutwaters span the River Esk. Lady Jane Seton, d1558 is credited as patron of works here, possibly to an older structure damaged in 1548. Other repairs are recorded in 1597 and at various times in the 17th century.

## MUTHILL   NN 875154    7km south of Muthill

Something still survives of the bridge over the Machany Water built in the 1420s by Michael Ochiltree, Dean (later bishop) of Dunblane, but it has been widened on both sides subsequent to being widened just on the downstream side.

*The Nungate Bridge at Haddington*

*Panel on Bannockburn Bridge*    *Panel on Dee Bridge, Aberdeen*    *Panel on Dairsie Bridge*

## NEWBATTLE    NT 332658 & 336666    11km SE of Edinburgh

The bridge carrying B703 over the South Esk to the SW of Newbattle Abbey has two slightly pointed arches divided by cutwaters with refuges, that to the NE being curved. Further downstream to the NE is the slightly decayed late 15th century Maiden Bridge with a single ribbed arch spanning 14.5m carrying just an estate track. It is named after Princess Margaret of England, who stayed at the abbey on her way to marry James IV.

## PEEBLES    NT 251402    On the south side of the centre of Peebles

The five segmental arches of the Tweed Bridge are of 15th century origin but the superstructure dates from widening in 1834 and 1897-1900. Additional arches of 1793 at the south end were replaced by a railway bridge c1865. A ramp with five round-headed arches now connects the bridge to Tweed Green on the NE side.

## PEEKIE    NO 560126    6km SE of St Andrews

A single semicircular arch carries a track over Peekie Water. It is said to have borne a panel dated 1522 connecting it with St Andrews cathedral-priory.

*The parapet at Musselburgh*    *Pencaitland Bridge*

*Stirling Bridge*

## PENCAITLAND    NT 442690    8km SW of Haddington

The main arch of this bridge over the North Tyne is supported on five ribs. A shield reset on the widened downstream south side bears the arms of the Sinclairs and a date thought to be 1510. A secondary arch on the west only has medieval work at the haunches. A north-facing cutwater divides the main arch from another to the east, which has been underfilled by a more recent arch.

## ROSSLYN    NT 275628    11km south of Edinburgh

The approach road to the castle set on a promontory above the River Esk crosses a man-made gap in a natural causeway. A bridge with arches supporting parapets with a wooden roadway between them replaced the original drawbridge by the end of the 15th century. The western arch was replaced by a segmental one in the late 16th century and a squinch was needed to turn the road round the NW corner. The roadway beneath all this also once had medieval bridges across the Esk. One abutment still remains.

## STIRLING    NS 797745    NE of the centre of Stirling

Beside the present road bridge the River Forth is crossed by a 15th century bridge of four semi-circular arches each with two chamfered arch-rings with the lower one recessed. The cobbled roadway is 4.5m wide and the pedestrian refuges originally rose up to crowstepped gables. They were changed in 1748, when the SW arch was rebuilt after being blown up in 1745, and gates at either end were removed.

Just north of the bridge, and on a different alignment, have been found piers of the bridge which existed in 1297, when William Wallace held the north bank in defiance of a larger force led by the Earl of Surrey. After his vanguard had crossed the bridge and been cut to pieces Surrey had the bridge severed to stop pursuit of his fleeing army.

## TULLIBODY    NS 854940    6km east of Stirling

In 1560 Kirkcaldy of Grange is said to have had an arch or two of this bridge over the River Devon tumbled to delay the advance of French troops on Stirling. The bridge had four arches "bowis" in 1616 when the Privy Council authorised the levying of tolls to pay for repairs. Further tolls were authorised in 1675 and 1681 after a report of the bridge then being broken down. On the east is a raised pathway bounded by rubble retaining walls rising to a parapet. Spanning the river are two obtusely pointed arches with triple arch-rings, massive ribs on the soffits, and heavily buttressed abutments, all probably early 16th century. To the west a length of late 17th century causeway appears to be a substitute for one of the original four arches. Then there is a ribbed arch which is almost semicircular over a sidechannel. The small round arches of the causeway further west, one with a hoodmould, are probably early 19th century.

*Tullibody Bridge*

## UPPER NORTH WATER or INGLISMALDIE NJ 653661 10km NW of Montrose

The bridge built by Bishop Gavin Dunbar c1525 over the River Esk has four wide arches each with several chamfered ribs, and has been widened on the upstream western side. The abutments have curious brackets between the springing of the ribs. The bridge is now bypassed by a new viaduct further upstream. See picture on page 93.

## OTHER BRIDGES PROBABLY EARLIER THAN c1600

AVON NS 733546  Three rebuilt segmental and ribbed arches each spanning 10m.
DALKEITH NT 339677  Just one abutment remains of the Cow Bridge over River Esk.
GLENLUCE NX 191573  One round and one segmental arch. Widened in 1838.
INVERKIP NS 223725  Single round arch over Kip Water. Probably widened.
LINTMILL NT 622249  Widened arch over Ale Water.
LORNTY NN 172465  18th century arch over Lornty Burn on 16th century abutments.
MINNIGAFF NX 412670  Queen Mary is alleged to have crossed this bridge in 1563.
RUIM  NO 270493  Round arched packhorse bridge shown on Pont maps of 1600.
RUTHVEN NO 290489  Ruined abutment below and to north of bridge of 1855.
STRATHAVON NS 710422  Two segmental arches over River Avon. Now bypassed.
URR NX 776677  Two semicircular arches. Datestone of 1580.

*Rosslyn: remains of river bridge*

*Maiden Bridge near Newbattle Abbey*

Balgownie

Aberdeen

Upper North Water

Brechin

Alyth

Muthill

Inner

Guardbridge

Dairsie

Peekie

Doune   Tullibody

Stirling

Bannockburn

Pencaitland

Abbey

East Linton

Cramond

Musselburgh   Haddington

Maidens

Rosslyn

Newbattle

Peebles

Twizell

Ford

Jedburgh

Ayr

Alloway

Dumfries

Abbey Bridge is described under Haddington
Twizell & Ford are described in the Northern England section

*SURVIVING PRE-REFORMATION BRIDGES IN SCOTLAND*

# A GLOSSARY OF TERMS

Abutment — Solid masonry placed to resist the thrust of a vault.
Arch-Ring — Ring of rough slabs or cut voussiors forming an arch.
Ashlar — Masonry of blocks cut to even faces and square edges
Centering — Wooden frame used to support an arch or vault under construction
Chamfer — Surface made by cutting off square angle at 45% to other surfaces
Clapper — A large, roughly rectangular slab of stone
Coping — A capping or covering to a wall.
Corbel — A projecting block of stone to support a beam or other stonework.
Cutwater — A triangular beak on a pier in a river to deflect the flow of water.
Four-centered arch — An arch with each side drawn from two compass points.
Hood-mould — Projecting moulding above an arch or lintel to throw off water.
Impost — A bracket on a wall to support one end of an arch.
Keystone — The central voussoir of an arch, usually emphasised in some way.
Lintel — A horizontal beam or stone across an opening.
Machicolation — A slot between corbels used to drop or fire missiles on assailants.
Metalled — Provided with a hard surface, originally stone slabs or cobbles.
Parapet — A wall for protection at any sudden drop.
Pilaster — A buttress of shallow projection.
Pontage — Tax on goods coming into a town to provide funds for bridgeworks
Portcullis — Grille of iron or wood shod with iron rising and falling in grooves.
Soffit — The underside of an arch.
Spandrel — Surface between side of an arch & adjacent horizontal and vertical
Squinch — Arch across the angle between two walls to carry upper walling.
Staddle — A subterranean support for a masonry pier.
Starling — A platform around the base of a pier to protect it from scouring.
String-course — Projecting horizontal band or moulding set in surface of a wall.
Voussoir — Wedge-shaped stone used in arch construction.

# FURTHER READING

A Guide to the Packhorse Bridges of England, Ernest Hinchcliffe, 1995
Ancient Bridges of the South of England, E. Jervoise, 1930
Ancient Bridges of the North of England, E. Jervoise, 1931
Ancient Bridges of Mid and Eastern England, E. Jervoise, 1932
Ancient Bridges of the Wales and Western England, E. Jervoise, 1936
An Encyclopedia of British Bridges, D.McFetrich, 2010
A Roman Timber Bridge at Aldwinkle, in Brittania VII (1976), pp39-72
Bridges of Bedfordshire, Angela Simco and Peter McKeague, 1997
Discovering Bridges, L. Metcalfe, 1970
Irish Stone Bridges, Peter O'Keeffe and T. Simington, 1991
Masonry Arch Bridges, J. Page, Transport Research Laboratory, 1993.
Medieval Bridges, Martin Cook, 1998
Monnow Bridge and Gate. M.L.J. Rowlands, 1994
Old Cornish Bridges and Streams, H. Coats & C. Henderson, 1928
Old Devon Bridges C Henderson & E Jervoise, 1938.
Structural Aspects of Medieval Timber Bridges, S.E.Rigold,Med Arch XIX1975, p48-91
Sutton Packhorse Bridge, Peter McKeague, Beds Archeology 18 (1988) pp64-80
The Bridges of Britain, E. de Mare, 1954
The Bridges of Lancashire & Yorkshire, M.Slack, 1986
The Bridges of Medieval England, D. Harriison, 2004
The Bridges of Wales, G. Breese, 2001

# INDEX OF BRIDGES